K A Longman

Tropical Trees: Rooting Cuttings

A Practical Manual

Illustrated by
R H F Wilson

Blaketon Hall
Exeter

In association with the
Commonwealth Secretariat

The jacket photograph is by courtesy of Brian Kerr.

A CIP catalogue record for this book
is available from the British Library.

Cased edition ISBN 0 907854 28 1

This international casebound edition published in 2002 by
Blaketon Hall Ltd, Unit 1, 26 Marsh Barton Road, Exeter,
Devon, EX2 8PN, United Kingdom.

A spiral-bound paperback edition is available from the
Commonwealth Secretariat – *Rooting Cuttings of Tropical Trees*
(ISBN 0 85092 394 8).

Printed in the United Kingdom by Bookcraft,
Midsomer Norton, Radstock, BA3 4BS.

PREFACE

One of the most challenging tasks facing the tropical world, now and into the 21st century, is to halt and reverse the loss of its trees. People in many countries are realising just how important it is that tropical trees are being cut down so much faster than they are being replaced. They recognise that, to a far greater extent than in the temperate zones, tropical trees are integral to the activities of local communities, governments and whole regions.

It once seemed there would always be plenty of trees growing naturally, but most tropical regions are now suffering from serious shortages and increasing problems. The loss of so many tropical trees arises for many reasons, which include increasing human populations, dubious advice and short-term planning. However, it is clear that there is now a widespread movement towards protecting and replanting trees.

Scientists and growers have been so preoccupied with research on a few tropical food plants that they have tended to neglect the trees which are so important in maintaining soil fertility. Similarly, the range of materials, implements, medicines, foodstuffs and fuel which trees produce were also overlooked. In addition, the important roles that trees play in modifying the human environment and in stabilising the local and global climates are also now more widely recognised.

To encourage everyone to plant trees is perhaps the single most useful step that we can take for the drier as well as the humid tropics. Although this sounds simple, there are in fact many potential pitfalls involved in choosing and growing good trees, planting them well and achieving successful establishment. In addition, because the focus has been on a few species, the very large number of useful indigenous trees found in tropical forests and savannas are mostly undomesticated and little-known.

The aim of this series, *Tropical Trees: Propagation and Planting Manuals*, is to try and meet the urgent need for practical help for small- and large-scale growers. Volume 1 deals with the Rooting of Cuttings of Tropical Trees, describing techniques that offer two important advantages over relying only upon planting seedlings. First, vegetative propagation allows new trees to be raised at any time, so that seed problems need not affect the choice of suitable planting stock. Second, as has been the case with several common farm crops, vegetative propagation can speed up domestication to allow urgently needed tree planting to be carried out with mixtures of superior selections.

Manual 1 describes general methods and approaches that apply to the great majority of the more than one hundred tropical tree species that have so far been tested. Much of this research formed components of tropical projects associated with the Institute of Terrestrial Ecology, near Edinburgh, Scotland. Translation of the Manual into Spanish and French is in progress, and it is clearly desirable that other languages be included. We hope that the Manuals will stimulate users to write further sheets, for instance about a particular tree species or specific local conditions. The author will also be happy to receive suggestions as to additions and improvements for future editions.

An informative Video programme, to be filmed largely in the tropics, is also being prepared on the same subject as Manual 1. Consisting of five interlinked sections, it complements the written sheets, and is intended both for specific training and for general educational purposes. Further details can be obtained from the Edinburgh Centre for Tropical Forests, Darwin Building, Kings Buildings, University of Edinburgh, EH9 3JU, Scotland, U.K.

I would like to thank my colleagues at the Institute of Terrestrial Ecology, and at many other tropical projects and organisations throughout the world for their help and advice while I was preparing this manual.

Alan Longman

Contents
Rooting Cuttings of Tropical Trees

Check-lists, sources and records

Introduction A1

why grow tropical trees?

Who are these Manuals meant for?

For ***anyone*** who is going to grow and plant tropical trees, including people working in:

(1) forestry departments;
(2) private forestry and timber firms;
(3) agriculture and horticulture departments;
(4) farming, including small-holdings;
(5) agroforestry and community tree planting projects;
(6) protecting sites from erosion by water or wind;
(7) the reclaiming of degraded land;
(8) school tree-planting and demonstration plots;
(9) parks and roadside planting;
(10) tree planting for research.

Are the Manuals only available in English?

No, they are being translated into Spanish and French, and it is hoped that arrangements may be made for other languages.

Why are they needed?

(a) much research information isn't reaching the growers of trees;
(b) a lot of tropical experience could be passed on to others;
(c) although some methods are quite well known, others are not;
(d) practical details of methods need to be explained, step by step;
(e) background ideas behind why something is done are also useful;
(f) such general tropical Manuals could encourage local publications;
(g) nowadays there are many fewer trees, but many more people;
(h) tree-planting is being recognised as very important;
(i) there's a better chance to-day to plant genetically improved trees;
(j) although growing trees is straightforward, a problem at only one of the many stages can easily prevent successful establishment.

What sorts of trees are the Manuals for?

Any kind of tropical tree.

But there are hundreds!

Yes there are. But these Manuals deal with the ***general principles*** of growing them. These apply to most species, grown for many different reasons, throughout the tropics. It is hoped that they may support and stimulate ***specific, local*** publications and translations.

Is tree planting particularly important in the tropics?

Yes, it is. Trees play vital roles in ***producing*** wood and other goods, ***protecting*** the site and ***providing*** the conditions needed for farming.

Is most tropical wood exported?

No, some timber goes for export, for making furniture, panelling, etc.
But *much* more is used locally for:

(1) firewood and charcoal;
(2) building houses, boats, fences, etc;
(3) making furniture and household implements;
(4) turning into paper and packaging;
(5) converting into plywood and particle board (see Manual 4).

What do trees produce besides timber?

Many different kinds of materials, sometimes more valuable than wood, that can be continually harvested without damaging the land, *including:*

(a) fruits, nuts, bush-meat and other human foods;
(b) medicines, spices, honey, etc;
(c) fodder and browse for domesticated animals;
(d) latex, resins, waxes and dyes;
(e) cane and vines for basket making and tying;
(f) leaves for wrapping food, etc;
(g) shading and potting soil for tree nurseries.

How do they protect the site?

(1) by not allowing the surface soil to **overheat**;
(2) by slowing its **drying out**;
(3) by preventing heavy rain-drops **washing away** fine particles, and even all the soil; similarly, stopping soil from being **blown away** by wind;
(4) by picking up and recycling **mineral nutrients** which would be lost;
(5) by discouraging **weeds** that thrive in open clearings;
(6) by providing food and shelter for **animals** and ***micro-organisms*** that help the plant community to function normally.

But surely one gets rid of the trees in order to farm!

Only some of them! Except on fertile land such as swamps, farm crop yields generally decline sharply if you do not keep (or plant) enough trees to **preserve** (or **replace**) 'open woodland' conditions. The cover needs to be sufficient to protect the soil, but not so dense that competition for light and rooting space is too great for the farm crops to grow well. Traditional farming methods have often been successful because they used:

(A) an **intimate mixture** of many different tree and farm crops; *and/or*
(B) a **fallow period** of 15-20 years under dense, regenerating forest or savanna.

Do trees have any effects farther away from the site?

Yes, in several different ways:

(a) making flooding less likely by reducing the rapid run-off of rain;
(b) discouraging mud-slides and the covering of lower land with eroded soil, and slowing the silting up of reservoirs and harbours;

(c) 'combing out' dust and particles of pollution from the air;
(d) moderating local climate, and perhaps increasing regional rainfall;
(e) helping to counteract 'global warming', which might alter the conditions for human societies throughout the world.

Wouldn't it all cost too much?

The Manuals describe techniques that do not require expensive equipment or materials. Replanting the large areas of former forest and savanna lands that have been degraded will certainly cost a lot, but not as much as losing the use of them altogether!

Why weren't these things thought of before?

(1) because there seemed to be unlimited numbers of trees;
(2) because they were not considered an important resource;
(3) because ecological science was not thought to be relevant;
(4) because local experience has sometimes been ignored:
(5) because advisors often thought in terms of the temperate zones;
(6) because short-term gain had replaced *sustained yield value*.

This page has been left blank for field notes.

Introduction A2
– *overcoming the problems of growing good plants*

What sort of problems may I need to overcome?

Many different things can hold up plant production, including:

(1) **Accidental hazards:** such as fire, unexpected storms, flooding or drought;
(2) **Organisational difficulties:** for example, lack of experienced staff, training opportunities, suitable workers, adequate supervision, regular finance, relevant tools, materials and information;
(3) **Biological problems:** including difficulties with seed supply and germination, poor rooting of cuttings, and unsatisfactory growth or death of planting stock.

Surely you can't do anything about accidents!

One can't entirely prevent the chance of damage, but the risks can definitely be lessened, for example by:

(a) thinking about and recognising what they are likely to be, ***before*** deciding where the nursery should be sited (Manual 3).
(examples would be to avoid exposed or low-lying sites, and those near plantations liable to fire, such as teak);
(b) planting shelter- or fire-belts;
(c) training staff and workers to react quickly and effectively in emergencies.

What about organisational difficulties?

These quite often hold up progress. Indeed, without reasonable and stable working conditions, little effective tree planting may result from years of effort, training and expenditure.

The first aim of management is therefore to provide the circumstances in which good work can be done; *particularly by:*

(1) communicating well;
(2) planning ahead; ***and***
(3) encouraging morale.

This can be harder to achieve in international projects if there is poor communication between the many organisations involved (see A 60).

What kinds of biological problems are important?

Many difficulties can come up in tree nurseries, any one of which can hinder or stop the production of sufficient, good planting stock, even when all the other aspects of the growing environment are perfectly satisfactory.

These manuals cover many of the important points. (See A 61 for a ***Check-list of problems in growing good trees***, together with suggested solutions.)

But aren't some of them insuperable?

Although certain tree species have unsolved difficulties that hinder their planting, many do not. The main problems:

(a) are simple, day-to-day matters;
(b) are being studied by scientists and foresters, or are already sufficiently understood;
(c) can generally be got around (A 61).

What sort of day-to-day matters are important?

(1) Choosing suitable ***sources*** for cuttings (A 3-13) and seeds (Manual 2);
(2) Planning the ***nursery*** well (Manual 3);
(3) Avoiding unnecessary ***stress*** on nursery plants, which can easily damage, weaken or kill them;
(4) Preparing for planting so that ***stress*** is again minimised (Manual 4);
(5) Looking after planted trees until they are well ***established*** (Manual 5).

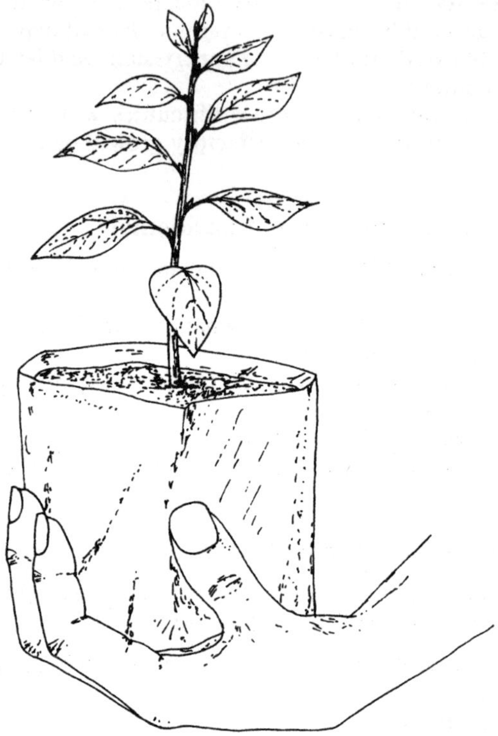

What kinds of stress can damage trees?

The commonest two types are:

(A) **Water stress,** caused by:
- (1) too ***little*** soil moisture;
- (2) too ***much*** soil moisture, preventing air reaching the roots freely;
- (3) a ***drying*** atmosphere *(sunnier, warmer, drier, and/or windier air)*; ***or***
- (4) a root system which is ***inadequate*** to replace the water lost from the shoot.

(B) **Stress from light levels,** either through:
- (1) too sudden removal of shade, exposing plants to bright sunlight (even for a short time); ***or***
- (2) Keeping plants in very dim light for several weeks.

Others include:

(C) **Temperature stress,** either:
- (1) too hot (above about 35°C); ***or***
- (2) too cool (below about 15°C).

(D) **Nutrient stress:** lack of *main* and/or *trace* elements (A 27, Manual 3).
(E) **Wind stress:** either mechanical damage, or through water stress.
(F) **Pests & diseases:** causing weakening, damage or death (A 52).

When are trees most easily damaged?

(a) When seeds are just germinating (Manual 2);
(b) When leafy cuttings are being handled (A 30-54);
(c) When plants have just been potted up (A 53);
(d) If unsuitable potting soils have been used (Manual 3);
(e) If conditions are ***suddenly*** made more stressful (A 54);
(f) If watering is not done regularly and well (A 51);
(g) If potted plants have been allowed to root freely through into the ground, because then they lose their main roots when moved (A 54);
(h) During the first week after planting (Manual 5).

But shouldn't trees be treated 'rough' if they are to survive?

No! This is a misunderstanding that has grown up. Tropical nursery stock need to be *hardened* **gradually,** not subjected to sudden shock treatment or abrupt changes in growing conditions.

Understanding how to grow good planting stock is the first step to successful establishment.

This page has been left blank for field notes.

Introduction A3
– *vegetative propagation*

What is vegetative propagation?

The production of new plants directly from vegetative parts of existing ones, not from seeds.

How is it done?

There are many different methods, including:

(1) **Rooting cuttings** – usually by encouraging roots to form on a piece of stem, so that it becomes an independent plant. *(Occasionally leaves can be stimulated to produce both shoots and roots.)*

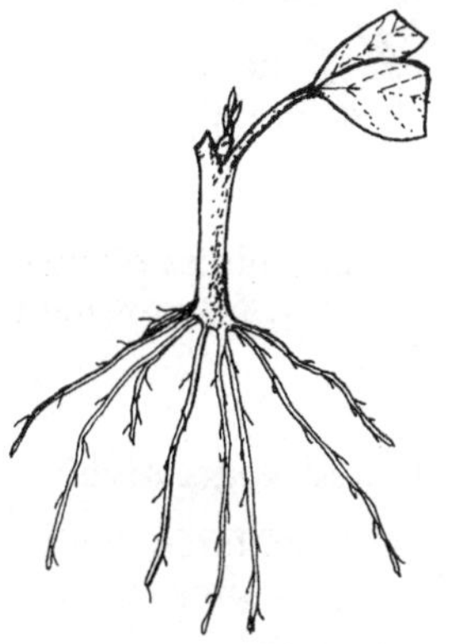

(2) **Grafting or Budding** – joining a piece of stem or a bud (the *'scion'*) to another plant (the *'rootstock'*), to provide root systems for detached pieces from a selected tree.

(3) **Planting shoot tubers or root tubers** – these storage organs are usually separated and replanted when they are not growing actively.

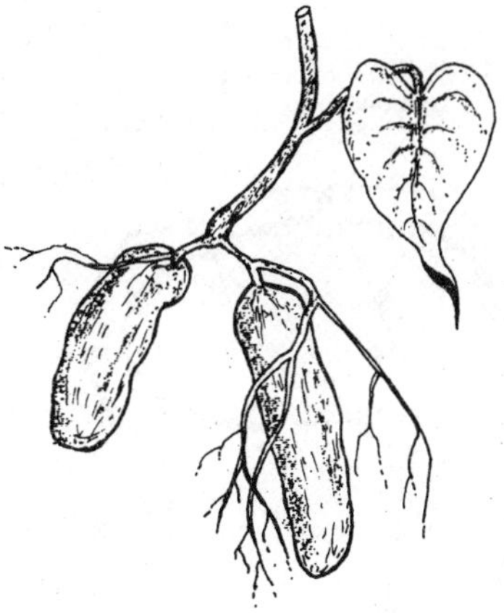

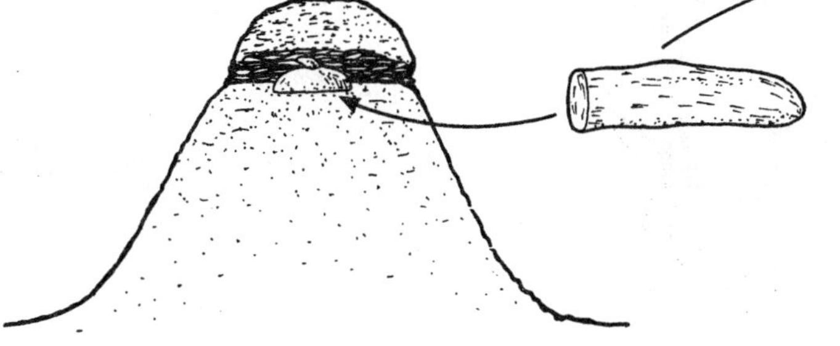

(4) **Taking suckers** – using shoots that have been produced on roots; either treating them as unrooted cuttings, or separating them with a piece of root attached.
(5) **Separating offsets** – dividing up plants that form clumps or sets of buds near to the ground level.

(6) **Micropropagation** – growing small pieces of tissue in sterile culture in a laboratory, stimulating many *plantlets* to form (A 5), growing these on; then potting, weaning and hardening them (A 53-54).

Which tree species can be propagated vegetatively?

(1) **Cuttings:** more than 80% of tropical forest trees so far tested can be rooted as ***leafy stem cuttings*** in low-technology 'poly-propagators' and/or under mist (A 30-32), grown on in a nursery and planted out just like seedlings. ***[This Manual deals primarily with rooting leafy stem cuttings.]***
A few will root from ***leafless stem cuttings*** placed directly into the ground, as is done with cassava (manioc) and several shade, hedging and ornamental plants (see A 4 for more details).
[Many *Begonia* species can be propagated from ***leaf cuttings***].
(2) **Grafting:** has been successful for example with *Cedrela, Cordia, Mangifera, Pinus, Tectona, Terminalia, Treculia, Triplochiton*, and is likely to be possible with most tropical trees. It is used especially for seed orchards (see Manual 2).

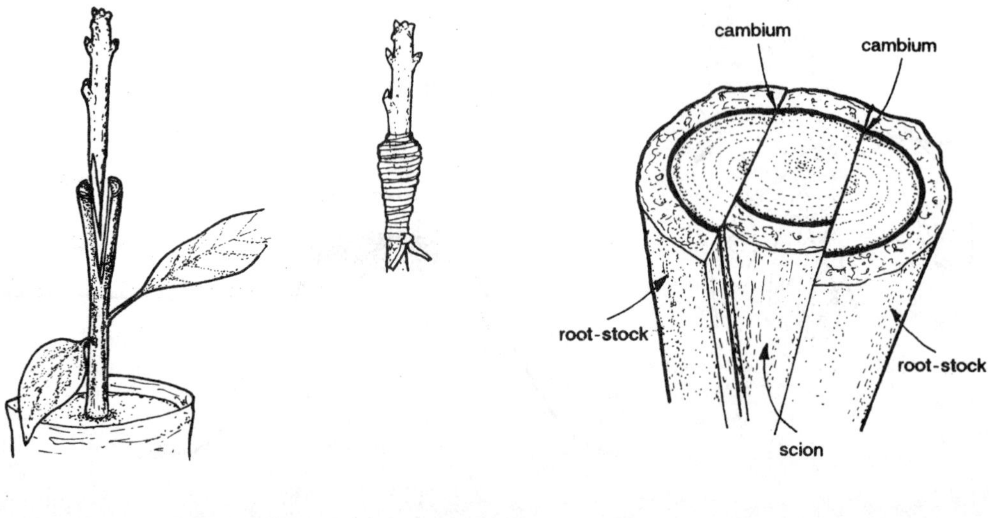

(3) **Tubers:** used for agricultural crops such as potato **(shoot tuber)**, and yam and cocoyam **(root tubers with a bud)**.
(4) **Suckers:** occur in *Acacia dealbata, A. melanoxylon, Chlorophora* spp., *Cinnamomum camphora, Cordia alliodora, Melia azedarach, Millingtonia, Ocotea usambarensis* and *Populus canescens*.

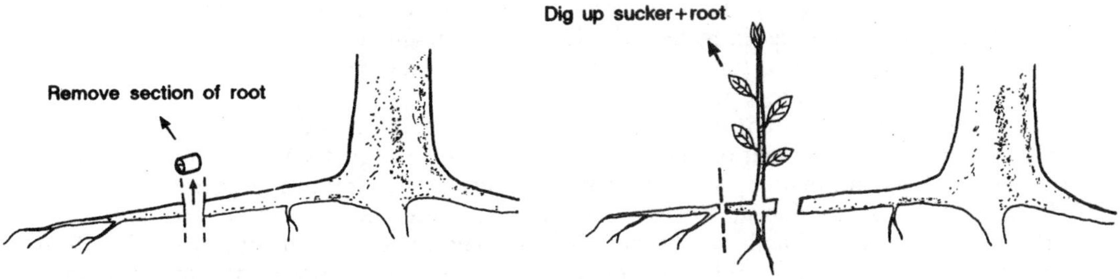

(5) **Offsets:** are used in farm crops such as plantain and banana.
(6) **Micropropagation:** has been successfully done in *Pinus* and oil-palm, is easy in *Nauclea diderrichii* and more difficult in *Khaya ivorensis* (see A 5).

What's the point of vegetative propagation?

There are two very important reasons why it is done:

(A) **To be free to grow a species** – many tropical trees ***flower and fruit rarely;*** and/or their seeds are ***not easy to store***.
By rooting cuttings, plenty of nursery plants can be obtained at any time, without having to rely on seeds.
Vegetative multiplication is also particularly valuable:
(1) if pests or diseases destroy most of the seed crop;
(2) when fruits or seeds are hard to collect, so the crop is small;
(3) with young material, before trees reach the flowering stage.

(B) **To get improved trees quickly** – most tree species ***have not undergone selection by mankind;*** and ***breeding is much slower*** than with annuals.
Vegetative propagation allows more rapid genetic improvement, because:
(1) trees with desirable, *inherited* characters can be used directly to produce improved planting stock (A 10-13);
(2) this direct selection captures more of the desired characters than is possible by seed;
(3) with each selected *clone* (A 11) existing as many plants, it is easier to check which characters are strongly inherited;
(4) clonal trials (Manuals 4 and 5) can provide a more rigorous test than a *progeny trial* of seedlings, and the results can be used at once.

Can't rapid tree improvement come through seed?

A good start on genetic improvement can be made fairly quickly through provenance trials and seed stands (Manual 2).
In a few species, seed orchards may give good results.
In many others further progress can be slow, because:

(a) there may be long delays before the selected tree flowers;
(b) selecting trees that have large flowering crowns could lead to ***negative*** selection when the aim is to produce a vegetative yield from the main stem;
(c) the characters selected for in the female (seed) tree may have been strongly influenced by environment, competition or chance;
(d) the male (pollen) parent is usually unknown, and may be inferior;
(e) the inherited tendencies of the parent trees are mixed during reproduction, so that less favourable combinations are transmitted to some of the seedling progeny;
(f) inbreeding depression may occur, caused by crossing between closely related trees;
(g) it can often be difficult to obtain large and regular supplies of improved seed from grafted trees in seed-orchards, though in a few species flowering can be reliably stimulated (Manual 2);
(h) there can be graft incompatibility between scion and rootstock.

How do I know that cuttings will do better than seedlings?

They will only have the potential to do so if they are genetically superior. However, there are several ways to make it more likely:

(1) By planting 10-30 clones together, not just one or two;
(2) By using selected clones (A 11-13), if possible those that have been through a clonal trial (Manuals 4 and 5);
(3) By selecting for those characteristics of stem form and branching habit that show up early in life;
(4) By making sure that cuttings come from a suitable shoot (A 6, 20);
(5) By ensuring that root systems are as good or better than those of seedlings currently being used;
(6) By labelling some clonal lines, to see which do best.

Are cuttings useful for anything else?

Yes, vegetatively propagated trees are an **important research tool**. Because they are genetically uniform, the effects of ***environment*** and ***internal changes*** within the tree can be separated from ***inherent*** effects.

Some misleading ideas about vegetatively propagated trees:

(a) "They won't produce a vertical stem" – ***only true when unsuitable shoots or 'mature' material are used (see A 6, 21), or inferior clones.***
(b) "Their root systems are no good" – ***only true if the cuttings were poorly rooted or had their roots badly restricted in a container. When rooting is stimulated with auxins (A 40), the root systems can be even better developed than those of seedlings.*** (Note that main vertical roots are seldom necessary for either (Manual 3).)
(c) "They are too expensive" – ***often said during the research and development stage, before it is sensible to compare either the costs of production or the benefits of species availability and genetic improvement.***
(d) "Their genetic diversity is too small" – ***only if very few clones are used. Successful clonal tree planting harnesses the useful uniformity of the selected clone within the safety of a diverse multi-clonal mixture.***

Everyone interested in growing tropical trees can now take advantage of the move away from just planting 'wild' trees.
Vegetative propagation offers the best prospect of growing much improved, 'domesticated' crops, as in agriculture.

Introduction A4
– *leafless and leafy cuttings*

Are there other names for these two types of *cutting?*

Leafless stem cuttings are sometimes called *hardwood* or *winter* cuttings; *stakes or poles*.
Leafy stem cuttings have been termed *softwood, greenwood, semi-hardwood, ripewood,* or *summer* cuttings.
These terms are confusing, and some are unsuitable for the tropics.

What's the difference between the two kinds?

Leafless cuttings are larger and of firmer wood. They do not dry up fast, and can survive in moist soil until roots are formed.
Leafy cuttings are smaller and are taken from softer shoots. Because they would dry out very quickly, they need to be kept in a humid *poly-propagator* or under *mist* (A 30-32) until they root and can take up sufficient water.

Why bother with leafy cuttings, then?

Because most species of trees can be rooted with leafy cuttings, but not so many will root from leafless cuttings.

Isn't it much simpler to root leafless cuttings?

Yes it is. Propagators are not required; pieces of stem are just pushed in, or put into a hole dug in the ground, preferably in moist, lightly shaded conditions.
If necessary, coarse sand might be worked into the soil to improve rooting; and simple, temporary shading could be placed over them in hot, dry weather.

How big should they be?

Sometimes 1-2 m poles, put in to mark out plots or planting positions, will take root and produce new shoots. This capacity is used to establish "live fences" and "live poles" to support climbing crop plants, for example in *Cassia siamea, Gliricidia sepium* and species of *Bombacopsis* and *Spondias*.
When a tall tree is to be grown, however, the ideal length is usually 15-40 cm, with a basal diameter of about 6-30 mm.

What kinds of woody plants will root as leafless stakes?

Some other examples are *Bosqueia angolensis, Ceiba pentandra, Debregeasia salicifolia, Ochroma lagopus, Pterocarpus indicus* and species of *Dracaena, Erythrina, Ficus, Vernonia* and bamboos; agricultural crops such as cassava (manioc) and sugar-cane, and several ornamental and hedging shrubs.

Gmelina arborea has been rooted as leafless cuttings, by using the shoots removed from 'stumped' planting stock, without their leaves.

Does it matter when I take leafless cuttings?

(1) **Adequate soil moisture:** if possible plant in the first half of the rainy season, or in a place that can easily be kept watered.

(2) **Inactive buds:** the cuttings should preferably not have actively flushing buds, as it is better for new shoot growth to follow rooting. Lengths of older wood are often used, rather than the softest, green shoots.
(Note: leafless cuttings of a few species can require cool storage, to make them ready to root well).

(3) **Leaves:** shoots that have already lost their leaves naturally are often best; if not, cut through the leaf-stalks rather than tearing them off, and leave a few very small leaves at the tip.

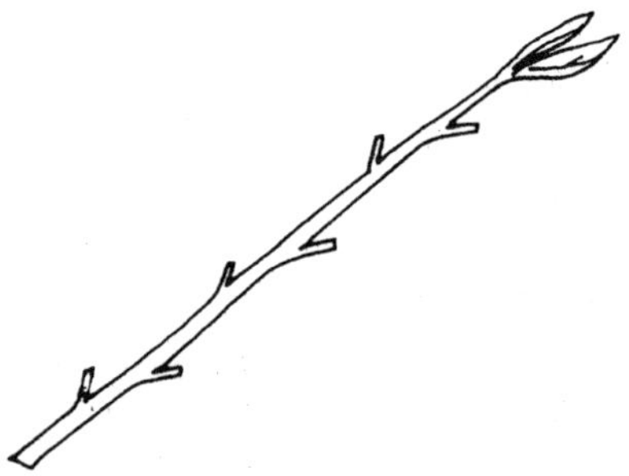

How should they be planted?

(A) **To grow a tall tree:**
Dig a hole, and plant the cutting with only about 3 cm above ground, including a live bud.

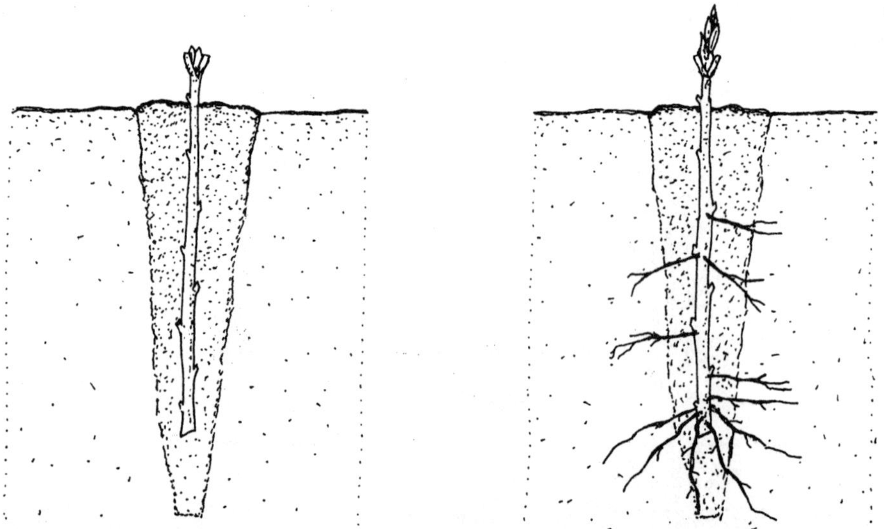

(B) **For shade trees or hedging:**
Point the stake and push it into the ground where it is to grow, leaving 0.5 to 1.5 m above ground. *(In hard ground use an iron bar.)*
Near-vertical planting is likely to produce a plant with a single shoot, whereas slanting or horizontal cuttings tend to form several shoots.

What about leafy cuttings?

Most of this Manual deals with rooting leafy stem cuttings. Sheets A 20-27 explain their production, A 30-36 deal with conditions for propagating them, and sheets A 40-54 cover how they should be handled.

This page has been left blank for field notes.

Introduction — micropropagation A 5

What is micropropagation?

One of the methods of vegetative propagation (A 3), which has to be done in a special laboratory.

How is it done?

Very small pieces of a tree are washed free of all bacteria or fungi, grown on a special *nutrient medium* in ***sterile culture***, and encouraged to produce many *plantlets*. These are then potted, grown on, weaned and hardened.

What are the main steps?

(1) **Setting up the tissue culture laboratory.** Two rooms without windows are needed, which can be kept very clean. One requires banks of fluorescent lights, and air-conditioning that can maintain a temperature of 25°C. The second, used for preparation, needs a *sterile-air bench*, and storage space. An *autoclave* is also required for sterilising the *media* on which the cultures will grow.

(2) **Preparing the media.** There are several different media which have given good results, and the basic chemicals can be obtained ready mixed (A 63). Distilled water and agar are added, together with specific quantities of the *plant growth regulators (hormones)* that determine whether shoots or roots are formed. ***[Note:* Cytokinins *tend to stimulate shoots, while* auxins *promote roots (A 40); but both are often needed.]***
After autoclaving – to kill spores of micro-organisms – the medium is poured into sterile tubes on the sterile-air bench, closed with cotton-wool, and allowed to set at an angle, producing a sloping surface. ***[Note: some heat-sensitive hormones must be added after heating, using a sterile syringe.]***

(3) **Taking the 'micro-cuttings'.** Small pieces of tissue are taken from the selected tree with a sharp blade, and throughly washed in a disinfectant such as *hydrogen peroxide* (H_2O_2) or a solution of *hypochlorite*. Success has usually been achieved with strongly juvenile material (A 6), such as parts of newly germinated seedlings, but small, active buds and even flowering parts have also been used.

(4) **Setting the 'micro-cuttings'.** Using the sterile-air bench and sterilised tools, a tube is opened and a single piece of tree tissue planted on the middle of the slope. A little sterile water is added, the tube closed again and placed under the lights.

(5) **Shoot multiplication.** The aim of the first stage is to encourage the culture to form many small shoots. This may take a few weeks, or longer.

(6) **Sub-culturing.** Once the shoots on the 'micro-stockplant' can be handled, they should be transferred to a fresh tube containing a medium that will promote rooting. If shoot multiplication is slow, the whole culture will need transferring to a new tube each month.
(7) **Rooting.** In the second stage, each shoot forms a root system.
(8) **Transfer of plantlets to soil.** The new plantlets need to be placed carefully into a sterilised potting soil in small pots, covered with a polythene bag and allowed to grow larger under the same conditions as the cultures. ***[Note: At this stage they are still very delicate, and need 'intensive care'.]***
(9) **Weaning.** The established plantlets need to be ***gradually*** accustomed to less humid conditions and somewhat brighter lighting (A 54). It is easy to lose them if this stage is done too rapidly, particularly when they are taken to a shaded nursery bed. However, if weaning is too slow the shoot may become drawn and weak.
(10) **Re-potting.** During the nursery part of weaning, transfer the plants carefully to larger pots, using un-sterilised soil. The inoculum for mycorrhizas or nodules can be added to the potting mixture (A 53).
(11) **Hardening.** For light-demanding species, the shading can be gradually reduced after the plants have been fully weaned (A 54).

Isn't all this very difficult to do?

There are usually problems in setting up the laboratory and in finding out what conditions are needed for a new species. Once routines have been worked out for a tree species, micropropagation can become more straightforward, but it is still time-consuming.

What's the point of it, then?

Micropropagation can sometimes give certain distinct advantages:

(a) **when rooting ordinary cuttings is difficult.** For example, oil-palm does not produce shoots suitable for cuttings, but selected clones are now produced by micropropagation.
(b) **for easy international exchange of selected clones,** free from pests and diseases. *(Note: viruses may still be present.)*
(c) **to eliminate viruses from a selected clone,** as has been done with cassava (manioc), for example.
(d) **if *adult* material from older selected trees can be *re-juvenated*** through micropropagation, this would be of great importance (A 10).

What are the disadvantages, compared with ordinary cuttings?

(1) **Propagation facilities** are more complex and expensive.
(2) **Knowledge of optimal conditions** is little known as yet, and the research for each species can take a long time;
(3) **Amount of work** is greater, especially in the early stages.
(4) **Staff and workers** need to be more highly trained;
(5) **Cultures can be lost** through electricity failure or rapid spread of contamination.

Where do problems happen most commonly?

(a) **In sterilising** the plant material;
(b) **In maintaining good growing conditions** for the cultures;
(c) **In obtaining shoot multiplication;**
(d) **In transferring plantlets** to potting soil;
(e) **In weaning and hardening** the small trees.

Introduction A 6
– *what makes a good cutting?*

Won't any tree of the right species give good cuttings?

No, because there is so much variability ***within*** each species in:
(1) **branching habit and stem form** (the individuals may vary from tall, straight, fine-branched trees to spreading, heavy-crowned bushes);
(2) **growth rate** (some may grow several times faster than others);
(3) **other desirable features of the tree,** such as fruit yields or timber quality; ***and***
(4) **rooting ability of the cuttings** (their potential could be anywhere from 0-100%).

What makes their shoots so variable?

(a) **Genetics,** the inherent differences that result from each individual tree getting a unique set of *genes* (genetic instructions) from its parents;
(b) **Internal changes** within the tree as it grows older and larger; ***and***
(c) **Environment,** all the climatic and site factors influencing the tree, and the effects on it of animals, *micro-organisms* and other plants.

Can all these effects be easily distinguished?

Usually no – it is generally difficult to tell just how much each is involved (although foresters often speculate). This is partly why it is much harder to draw firm conclusions from research on trees, compared with studies on most agricultural crop plants.
Sometimes yes – particularly if ***clonal*** **rooted cuttings** are used in experiments in place of seedlings. The effects of genetic differences can then be largely separated out and studied, which in turn makes it easier to find out what influences the other two are having.

What can I do about genetics?

Take your cuttings from trees that are of good:
(1) **provenances** (or local races – A 10): already shown to be suited to the proposed planting site (Manuals 4 and 5);
(2) **progenies** (Manual 2): coming from parents that have produced good trees before;
(3) **clones** (A 11): of tested clones, if available; or from individual trees that are healthy and vigorous, and are straight with fine branches (A 12), or have other desirable features.

Genetic selection is covered in more detail on A 10-13 and in Manual 2.

Won't any shoot on a selected tree do as a cutting?

No; partly because **internal changes** will have made some of them unsuitable for vegetative propagation, for example:
(a) **Rapid loss of rooting ability with age.** Most tree species root easily from leafy cuttings taken from young seedlings, but not when these are taken from the crowns of saplings or mature trees;
(b) **Branches unsuitable for propagation.** In many tropical trees, the branches have a different structure from the main stems, and after rooting they give plants which ***may not show a normal growth habit;***

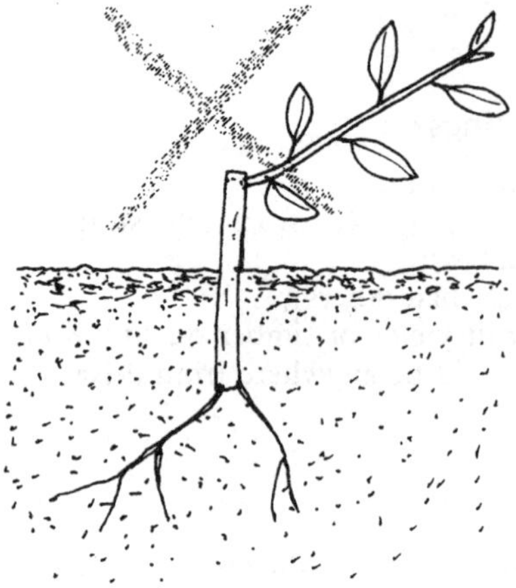

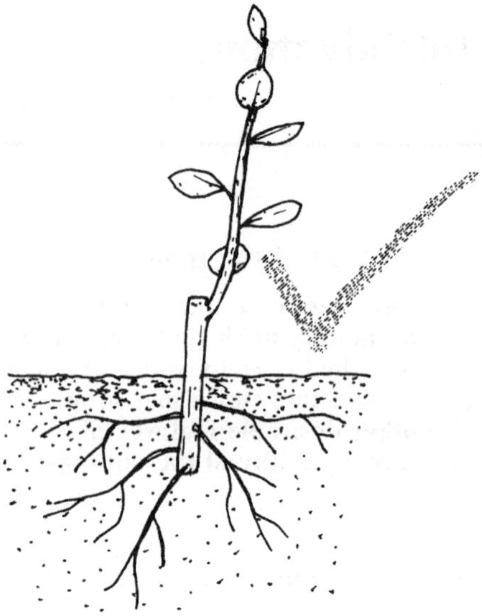

(c) **Vigour of growth.** Leafy cuttings generally root best when the shoots were growing at a moderate rate before they were taken (A 42).
Take leafless cuttings when the buds are ***not*** making new growth (A 4).

But all these things will stop me getting enough cuttings!

This may be a problem at first, but can be overcome by:

(A) **Using nursery seedlings** as a source of cuttings to start with;
(B) **Using managed *stockplants*** to produce plenty of the right sort of shoot (see sheets A 20-27 for more details).

How are stockplants managed?

Here are two ways:

(1) **By felling a selected tree, and repeatedly harvesting the coppice shoots it produces.** Coppice shoots from within about 1 m of the ground are usually juvenile (like shoots from seedlings), often producing vigorous shoots with a main stem structure that root easily. Coppice stumps need frequent harvesting and pruning to prevent tall shoots and those with branch structure developing (A 21).

(2) **By planting *clonal* cuttings as stockplants.** These also need frequent harvesting and/ or pruning to persuade them to produce plenty of the right type of shoot (A 25). Alternatively, they can be grown for 1-3 years and then coppiced.

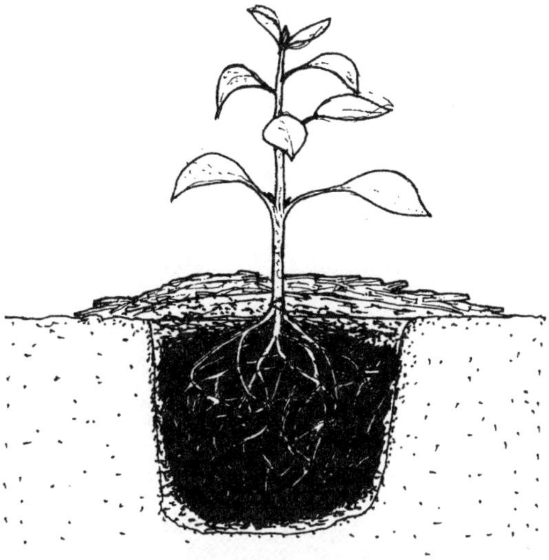

Can I change the environment to produce better cuttings?

Yes, for the stockplant environment makes quite a difference to how the cuttings root later on. For example:

(a) **Shading:** moderate shading of stockplants greatly increases the subsequent rooting of *Triplochiton scleroxylon.* The light filtering through green leaves is best, so it is a good idea to interplant with a leguminous shrub such as *Leucaena,* which has many small leaflets giving a suitable kind of shading (see A 24). *(Too much shade decreases subsequent rooting.)*

(b) **Watering:** During dry weather stockplants may grow slowly or stop producing shoots altogether. If they can be watered (A 26), it may be possible to harvest cuttings all the year round, and they are more likely to be of a suitable type.

(c) **Mulch and fertilisers:** mulching the stockplants (Manual 5) can help them establish quickly, grow well and produce plenty of good cuttings. Adding NPK fertilisers (A 27) can help in the same way, and also replaces minerals removed by harvesting shoots, or washed away in rainfall and watering. Since fertilisers can increase or decrease the subsequent rooting of cuttings, use them in small quantities after harvesting, or do experiments with your species (A 45).

Are there some kinds of tree without all these problems?

Some species that root easily from leafless cuttings (A 4) have fewer problems, and cypresses for example often produce large numbers of easy-rooting, vertically-growing cuttings without special attention.

However, most tropical trees need stockplant management to gain the benefits of vegetative multiplication and clonal selection.

This page has been left blank for field notes.

Genetic selection – *introduction* A 10

What is meant by genetic selection of trees?

Selection means choosing particular trees with desirable features as your source of plants, in preference to other trees or unselected origins.

Isn't that easy?

No, because a tree will be like it is for a variety of reasons (A 3, A 6):

(a) **genetic factors:** that is, inherent characteristics passed on to it from both the male and the female parent trees;
(b) **internal factors,** e.g. the place the cutting originated from in the stockplant, or how long it is since the original stockplant germinated;
(c) **environmental factors,** such as shading, soil type or water stress; and competition with other plants, or heavy grazing by animals; ***and***
(d) **chance factors,** for example damage that just happened to occur to that tree, rather than to the next one.

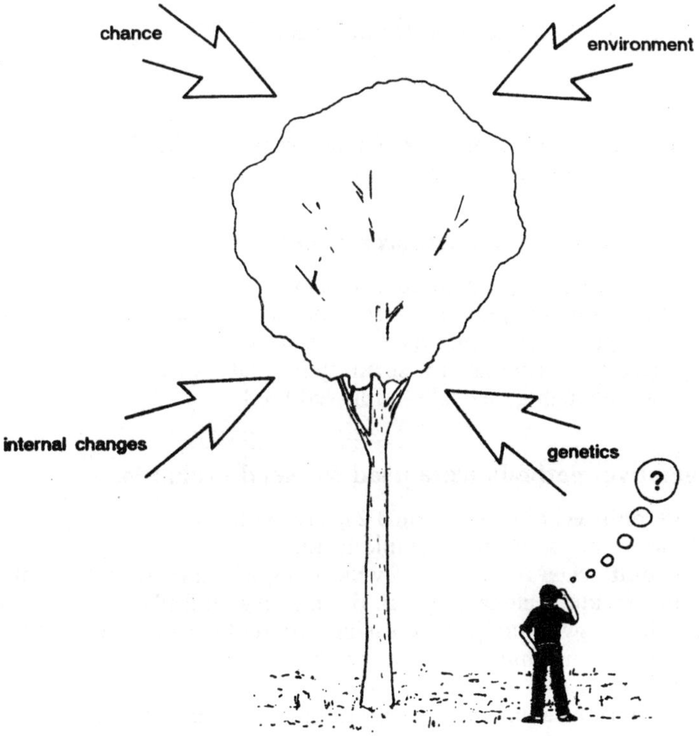

Only features influenced by **(a)** will be passed on, so genetic selection means ***learning how to choose trees so that desirable characters which are strongly inherited are favoured.***

How much genetic improvement is possible?

A great deal. This is because most tree species:

(1) **contain a lot of inherited diversity,** so there is plenty of potential; ***and***

(2) like wild animals, **have not yet been 'domesticated',** so we are still near the beginning of the process.

Why hasn't more been done, then?

For four main reasons. Because:

(A) it is **more difficult** than with annual crop plants, as trees are large, variable and *long-lived;*

(B) results often don't show up for **a long time;**

(C) promising approaches were **overlooked** for decades; ***and***

(D) **fewer resources** have been put into research on trees.

Is it more difficult still with tropical trees?

Yes, because they:

(1) have been even less well studied;

(2) are of hundreds of different species;

(3) often flower irregularly, high in the crowns;

(4) may give good seed crops only at intervals;

(5) can belong to various different genetic systems;

(6) generally produce quite variable progeny.

What approaches were overlooked?

The rooting of 'juvenile' shoots as cuttings, for instance.

Why did this happen?

Because the first forest geneticists concentrated on improvement through seed, instead of combining the two ways of producing planting stock.

Well, most trees are grown from seed, aren't they?

Yes, and a good start to tree improvement comes from:

(a) choosing seed of a suitable *provenance* or *local race* (Manual 2);

(b) taking seed from parent trees that have given promising *progeny.*

However, further progress is generally by small steps that can take many years, and may sometimes fail to provide enough genetically improved seed.

But I thought vegetative methods were used for seed orchards.

Yes, but these are different techniques (Manual 2), involving:

(1) selection of large trees showing desirable features;

(2) collection of **'adult'** *scions* from their crowns, and grafting these on to seedling *rootstocks;*

(3) growing on of the adult, clonal *grafts* in the nursery, and planting them out;

(4) waiting until they flower, and then crossing different parent clones with each other;

(5) growing the crosses in a nursery, planting them out in progeny trials, and observing and measuring them;

(6) removing the inferior parent trees, re-grafting the better ones, and waiting until they flower and yield seed crops.

What selection techniques are used when rooting cuttings?

They are simpler and quicker, for you can use the selected trees ***directly.*** Such systems have been used for decades for temperate-zone plantations of *Populus,* and over centuries for

Cryptomeria. Using similar methods, tropical farmers grow chosen strains of cassava (manioc), bananas, yams, etc. (A 3). Besides being simpler, several stages can be done at once:

(a) rooting cuttings from the best available seedlings for a start;
(b) selecting sizeable trees showing desirable features, felling them and using the coppice shoots as clonal stockplants (A 12, A 21);
(c) applying *early tests* to seedlings, and choosing the better ones (A 13);
(d) rooting cuttings from b & c, and testing them in clonal trials;
(e) planting stockplants of the tested clones, and managing them to produce many shoots that will make good cuttings (A 6, A 20-27);
(f) rooting the selected clonal cuttings in large numbers, and using them as planting stock.

Why has less effort been put into tropical trees?

Because :

(1) their ***multiple*** value wasn't generally appreciated;
(2) there usually isn't a quick return on expenditure;
(3) logging has been separated from afforestation;
(4) forest science was disconnected from forestry;
(5) genetic improvement seemed too slow and difficult;
(6) forestry staff were often unaware of the potential of clonal tree planting, or had been prejudiced against it.

But everyone knows that tropical trees are very important!

Yes, we are learning fast that they ***produce*** hundreds of things that people need; ***provide*** the conditions needed for agriculture; and ***protect*** the local soil, the regional climate and the world's temperature (A 1).

This page has been left blank for field notes.

Genetic selection A 11

– *clones*

What is a clone?

All the vegetatively produced plants arising from a single tree of seedling origin.

What's so special about such plants?

They are all very similar to each other, and can look almost like a set of 'photocopies'. However, like any tree, their growth is also affected by environment, competition and chance damage.

Why are they so similar?

Because, although they may grow in different environments, they have exactly the same genetic make-up, like identical twins. ***[Note: very occasionally, a small but permanent genetic change can occur in a member of a clone, and this may be passed on in cuttings taken from it.]***

How different is one clone from others in the same species?

It can be completely different. For instance, some clones grow into tall, straight trees, while others are crooked and forked. In *Triplochiton scleroxylon,* certain clones can even be bushy, with several main stems or heavy branches.

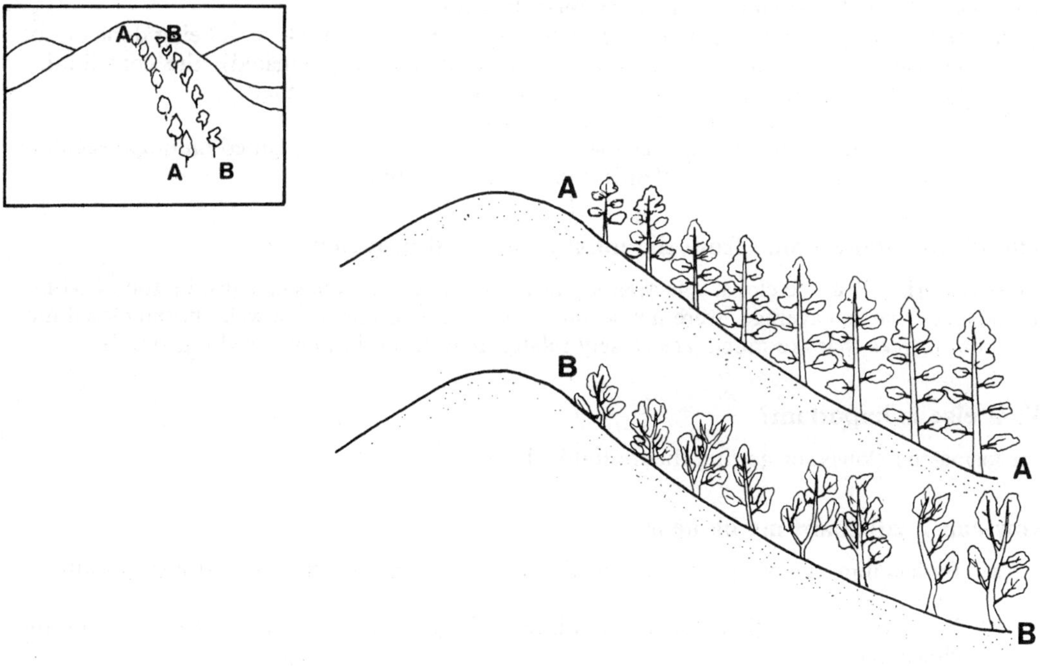

Will I get greater yields of produce by planting good clones?

When clones have been selected for early, regular, heavy fruit production, as has been done, for instance, by grafting *Treculia africana* in Nigeria, the clones can be expected to out-produce trees raised from seed, just like selected rubber, oil-palm or mango clones.

Is the same true for yields of wood?

If clones have been selected for stem straightness and favourable branching habit, they are likely to show a considerable improvement in these characters over trees raised from seed.
Greater height and diameter growth can also be expected, although these are more strongly influenced by environment and competition.

Should I plant a single, highly productive clone?

No! A mixture of 10-30 good clones is recommended for tree crops that take many years or decades to mature. Stands in which every tree had the same inherent characters would be too risky, for that clone might prove susceptible to drought, pests or disease, or be found to have other unwanted features.

Doesn't 10-30 clones still reduce genetic variation too much?

Only for ***seed stands,*** which will provide seeds for later planting, where greater genetic diversity is important to avoid inbreeding; and for ***gene banks,*** which are storehouses of variability for future tree improvement (Manual 2).

How do I know that the clones will be good ones?

Everyone recognises the (genetic) differences between clones of plantain, banana, mango, etc. Clones of trees have only recently been selected, so it will take longer to get to know them. However, forest scientists have shown that some 'domesticated' *Eucalyptus* and *Triplochiton* clones have the potential to outgrow and outproduce by far the 'wild', seedling trees.

Will we have to wait long to get supplies of these new clones?

They will need to be multiplied and tested first (Manual 4). However, the waiting time will be much less than with tree improvement by seed, because:

(1) the clones ***themselves*** are being directly tested, rather than selection being done on the ***average*** past performance of the progeny (often open-pollinated), the contributing parents in a seed-stand, or the provenance or local race (Manual 2);
(2) procedures can be shortened (A 12-13); ***and***
(3) clonal material (for instance as marked lines), could be interplanted **amongst** seedling stock, before supplies are plentiful or testing is completed.

Will all the clones come from the most successful provenance?

Not necessarily. Because clonal selection is more precise, good clones can also be found within provenances that, on average, were not so successful. This means that a well-chosen clonal mix can contain trees that are much less closely related than those in many seedling stands.

What else is important?

Not losing the clones, or getting them muddled up.

How can I safeguard my clones?

(a) By labelling the original source plant, stockplants and batches of cuttings carefully (A 12, 64, 66);
(b) By planting stockplants in several places, with a shrub such as *Leucaena* between the clones (A 24);
(c) By keeping some plants of each clone in pots (A 23);
(d) By watering carefully (A 26, 51) and inspecting frequently.

Genetic selection – *choosing good trees* A 12

Is it difficult to select trees?

Not once you have "got your eye in" with a particular species.
Remember that the desired features will be different when trees are being grown primarily for fruit, latex, shade or ground cover, rather than main stem wood.

What size of tree should I select?

Trees from about 3-30 m tall are most suitable *(see A 13 for selection tests for small trees).*

Is it easier in plantations?

Yes, much simpler, because one is choosing between trees of about the same age. However, natural trees can be included if they occur.

How do I know the tree's good points will be passed on?

Only the inherited features will be passed on (A 3, A 10). Fortunately, however, several important features are ***strongly inherited,*** so that there is a good chance that selection will be effective.

What are they?

For example, ***stem form, branching habit, wood structure, tolerance to disease*** and ***early flowering*** are usually highly inherited.

How do I select for good stem form?

(a) Find an accessible stand which has some promising trees in it;
(b) Walk through it with a colleague, looking for stems that are:
  (1) **straight,** or nearly so;
  (2) **unforked,** *persist* up into the crown and lack multiple stems;
  (3) **cylindrical,** without much fluting, heavy buttresses or signs of spiral grain;
(c) Give a score (from 0 – 5) to any promising trees *(Note: look at trees from several different angles – see A 65);*
(d) Reject any tree that one or both of you are not happy about;
(e) Paint a number at the base of selected trees, and the same number at about breast height (1.30 m); *(this will be its **clone number** – see A 11);*
(f) Make a sketch map so that you can find all the trees again.

Should I score all species the same?

No, because the natural growth habit is much better in some species than in others. For example, *Terminalia* trees are often straight, so your selection standard should be high. On the other hand, *Lovoa trichilioides* is often crooked and forked, and *Gmelina* heavy-branched (or sometimes with multiple stems), so here your standard may have to be less strict.

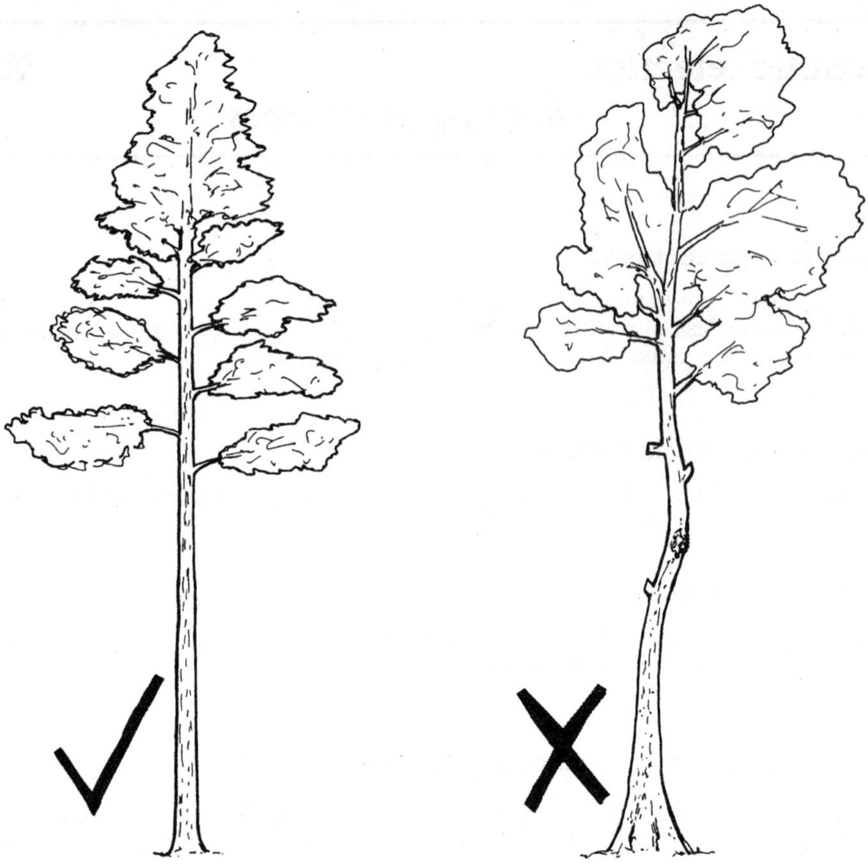

What about branching habit?

Score this ***at the same*** time as selecting for stem form, and look for:

(1) **fine** rather than heavy branches;
(2) **flat** rather than ascending or descending branch angle;
(3) **regular** rather than one-sided crowns;
(4) **deep** rather than shallow live crowns;
(5) dead branches **lost** rather than retained.

How about the size of the tree?

Yes, it is worth recording the diameter at breast height, and perhaps the height. However, as size depends so much on patches of different soils, competition and chance, give less weight to it.

But isn't a measurement more reliable than a score?

One ***can*** easily check that a measurement is accurate, whereas a score always depends partly on the opinions of those giving it. However, the important features like stem form and branching habit are hard to measure, but easy to see and agree on as an approximate score.

How can wood quality be assessed?

This cannot easily be done in the field, although it might be possible to send numbered wood specimens for testing when selected trees are felled to stimulate coppice shoots for propagation as cuttings (A 21).

Can disease resistance be selected for?

If disease is present in the stand, or there are signs of previous damage, then healthy, undamaged trees should be chosen. Otherwise, testing for tolerance to disease is best done as a clonal test at a later stage (Manual 4).

What about early flowering?

In trees giving useful fruits, select for:

(a) early flowering
(b) regular flowering
(c) heavy flowering
(d) fruit quality

Fruit set is less likely to be strongly inherited.
In trees grown for main stem wood, a – c may be ***undesirable*** (Manual 2).

Isn't it a pity to fell the best trees?

In one way, it is. However, until research workers find reliable ways of stimulating coppice shoots on intact trees, this method gives us the best chance of 're-growing' selected trees in large numbers. Coppicing lets us regain the juvenile stage, start new clones and make many trees of each.
After further testing, some of these clones may turn out to have outstanding combinations of desirable, inherited features that will greatly improve the quality of later plantings.

Does it matter where the trees are selected?

It is certainly true that easy access to the coppiced stumps, and nearness to the propagation nursery are both useful. However, it is **much more important** to search for good trees, even in remote areas. This is often where trees with favourable characteristics still exist, and you also need to include genes from different sites and parts of the country, to sample the available range of genetic variation.

This page has been left blank for field notes.

Genetic selection A 13
– *tests for small trees*

What are these selection tests for?

To find out which are the better and which the less good trees.

But surely you can't tell which trees are good till later on!

Some features (such as branching habit) often show up during the first few years; others (like wood structure) can only be found out later on.

How can the tests help?

By showing at an even earlier stage, *while the plants are still in the nursery,* what the inherent branching habit is going to be like, for example.

How is this done?

In *Cedrela odorata* and *Triplochiton scleroxylon,* for instance, the tip of the shoot is removed when nursery plants are about 20 – 30 cm tall.

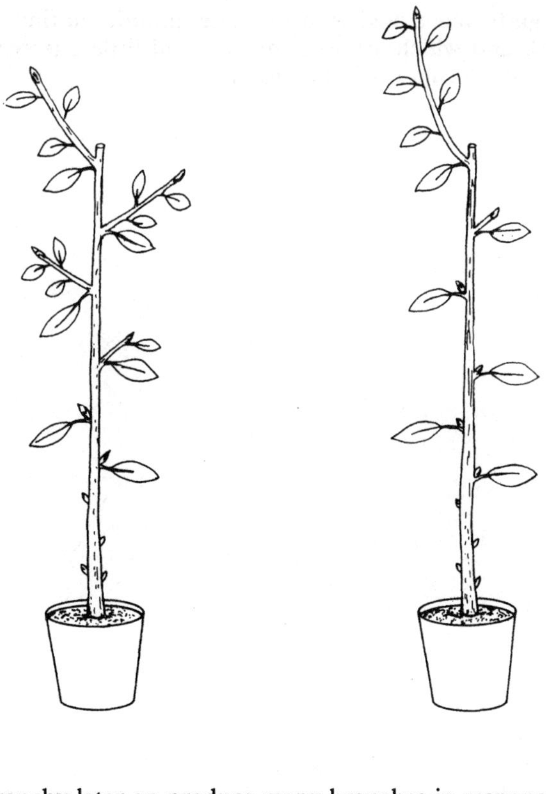

Trees which will be branchy later on produce many branches in response; whereas those which will put more into a **single main stem** produce only a few, one of which quickly replaces the lost leading shoot.

Do growing conditions matter for the test?

Yes, it is important that all the trees:

(1) are about the same height to begin with;
(2) are growing actively;
(3) were potted at the same time, with the same potting mixture, into similar size pots, and none have rooted through into the ground;
(4) are receiving similar lighting, watering and fertilisers.

How soon can I tell which are the best?

In *Triplochiton*, the outgrowth of branches reaches a peak in 2 weeks. Strongly dominant replacement leaders may form during weeks 3 – 6.

How can I use the selected plants?

If the test is done on a batch of:

(a) **seedlings** – discard the branchiest, using the best to start new clones (A 11) or as stockplants for bulk cutting production (A 20);
(b) **clones of rooted cuttings** – use the information to build up and confirm knowledge about the features of the different clones. Plant the best ones in a stockplant area for clone multiplication (A 20).

Can I use this test for other trees?

Research is going on to develop a test suitable for different species.

What else can be tested for when the plants are still small?

Sometimes small trees can be inoculated with a disease, in order to find out which of them are badly damaged or killed, and which are tolerant to it and little affected. Clones can then be started from those best able to cope with the disease.

Stockplant management A 20
– *introduction to vegetative multiplication*

What is vegetative multiplication?

Getting enough planting stock, but not directly from seeds (A 3).

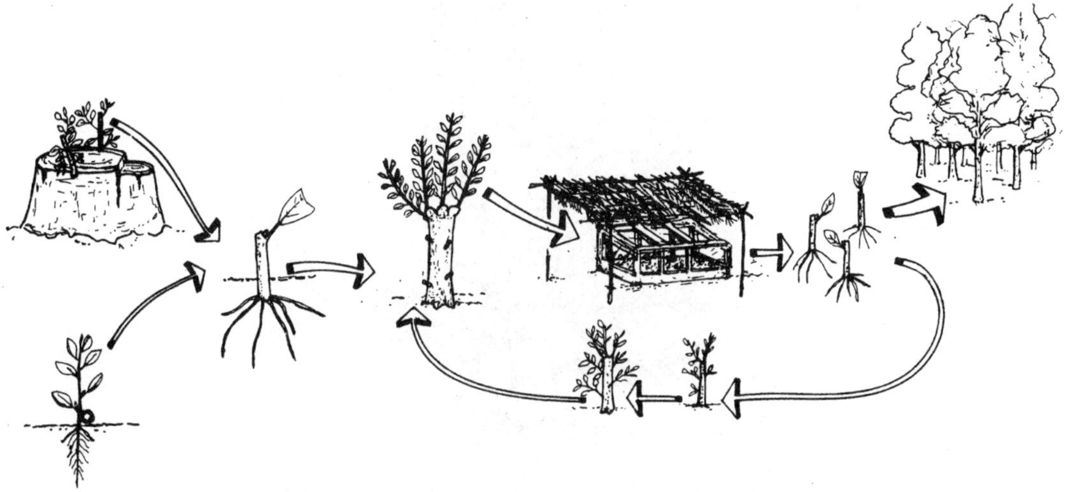

What's the advantage of that?

(A) Trees with desirable, inherited features can be used ***directly*** to produce improved planting stock (A 10-13).
(B) Plants can be produced each year in species with seed problems (A 61).

How do I set about it?

By establishing ***stockplants*** that will supply enough of the right type (A 6) of cuttings for rooting, by:
(a) planting and managing a stockplant area with the best available clones (A 11); ***and***
(b) felling and coppicing selected larger trees (A 12).

What are the snags?

Here are some common ones:
Problem 1 – cuttings are not of the right kind;
Problem 2 – not enough cuttings are produced;
Problem 3 – shoots are only available at one time of year;
Problem 4 – cuttings don't root well;
Problem 5 – coppiced stumps stop producing;
Problem 6 – the planted stockplants get older, producing poor shoots.

How do I get over these difficulties?

Problem 1 – prune stockplants to encourage good shoots (A 6, 21, 25).
Problem 2 – plant plenty of rows of each clone; improve fertility of soil (A 22, 27).
Problem 3 – water during the dry season (A 26).

Problem 4 – cut back (or coppice) stockplants (A 25); introduce shading (A 24); try other clones; improve propagation conditions (A 30 – 36).
Problem 5 – try watering in the dry season (A 26), or adding mulch or a little fertiliser (A 27); don't rely only on the stump, but also plant the clone in a stockplant area (A 22) and keep some plants in large containers (A 23).
Problem 6 – try re-coppicing or heavy pruning (A 25); if necessary plant a new set of stockplants raised from *juvenile* cuttings (A 6, 21).

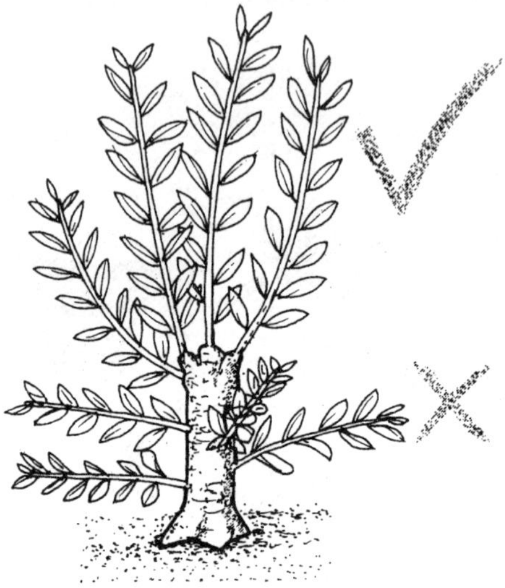

Won't numbers always be low at first?

Often yes, because young seedlings or cuttings used as stockplants don't yield many cuttings, so there are few plants of each clone in the early stages.

How can I speed up multiplication?

(a) Plant most of the early batches of rooted cuttings as stockplants;
(b) Use coppice stumps (A 21) as these can usually produce many cuttings repeatedly;
(c) Before selected clones are available, accept untested seedlings as stockplants for bulk production of unimproved cuttings;
(d) Later on, plant many stockplants of the most promising clones.

Are there any general hints for stockplant management?

(a) Plant stockplants in a place that is easy to get to;
(b) Interplant with a shrub such as *Leucaena* so that cuttings taken from different clones will not get mixed up (A 11, 22);
(c) Arrange a reliable watering system, especially if planting and harvesting are to continue during the dry season (A 26);
(d) Visit them ***regularly each week***, to see whether they are ready yet for harvesting cuttings, and to check for any damage, weeds or pests and diseases (A 52);
(e) Don't let the stockplants grow tall;
(f) Before clones have been selected, avoid taking most of the cuttings from the bushiest stockplants. This could turn out to be ***negative*** selection towards branchier planting stock!

Stockplant management – *managing coppice stumps* A 21

What are coppice stumps needed for?

They provide a valuable source of *juvenile* shoots to take as cuttings, especially useful when starting clones from selected trees (A 3, A 11-12).

Do all trees form coppice shoots when felled?

Most broad-leaved trees coppice, but only a few conifers do so.

If a species coppices, does every tree do so?

Clonal differences were found when 15 year old *Triplochiton scleroxylon* trees in Cameroon were felled at 4 different times of year. Most stumps produced a lot of vigorous coppice, some a few smaller shoots, while a few did not coppice. Six 25 year old *Lovoa trichilioides* produced hardly any coppice shoots, so it is best to do a few preliminary test fellings.

Will trees of any age coppice?

In general yes, but sometimes old trees may fail to do so, and very young plants may not have a large enough root system to coppice repeatedly.

At what height should the tree be felled?

At about 0.1 – 1.0 m above ground level.
This part of the trunk is most likely to produce juvenile coppice shoots that grow vertically and are easily rooted.

Where do the shoots grow out from?

From small inactive or newly-formed buds; either in the bark or on the cut surface (in wound tissue between the bark and wood).

How long does it take for coppice shoots to form?

The *Triplochiton scleroxylon* shoots were ready for harvesting in 1½ – 3 months, at any time of year.

Can I keep coming back to the same stumps?

Yes, provided that they are managed properly. Important points include leaving enough green leaves to keep the stump alive, maintaining moderate shade (see A 24) and pruning to stop tall, thick shoots and unsuitable branches from forming (see A 25).

How long will the stumps yield cuttings?

This depends partly on things you can't influence, like the **species** and the **clone**. But it can be extended to years by:

(a) felling with the top of the stump slightly sloping, so that rain runs off rather than encouraging it to rot;
(b) good management (as described here);
(c) always leaving some basal leaves behind when harvesting and pruning;
(d) adding a little fertiliser occasionally (A 27);
(e) watering during long dry spells (A 26).

Why do the stumps need managing?

So that they continue to produce plenty of cuttings of the right kind for easy-to-root, upright-growing cuttings (A 6).

What is the first step in good management?

Preventing the coppice shoots from getting too tall, or producing a lot of branches instead of vertical main stems.

How do I do that?

(a) By harvesting cuttings frequently, rather than occasionally;
(b) By often pruning back taller, thicker stems, leaving 2-3 basal leaves;
(c) By pruning to encourage shoots with main stem characteristics (A 6, A 25).

What else is important?

(A) **Light** – felling the tree usually leaves a gap amongst the crowns in the stand that gives the right ***amount*** and ***kind*** of light to the coppice stump below (A 24). However, it may become too shady because of:

(1) displacement of woody climbers during felling;
(2) increased growth of saplings, vines, undergrowth and weed species;
(3) rapid closing of the *crown gap* by new foliage.

In cases 1 & 2, give some more light (not too much) by cutting back on the shading plants as needed; for 3, maybe fell a nearby tree.

(B) **Protection** – against various kinds of damage:

(1) cut **vines** before they twine and choke the shoots;
(2) remove **insects** that are starting to eat or make nests in or near the coppice shoots – if necessary spray (A 52);
(3) spray if there are signs of damaging **disease**, or paint the cut stump soon after felling;
(4) make a strong fence if **animals** like hens, goats or antelopes are in the neighbourhood, or if necessary use *pollarding* (cutting stump above the animals' reach).
(Note: the cuttings from pollarded trees may not be juvenile.)

This page has been left blank for field notes.

Stockplant management – *planting stockplants* A 22

Why do I need to plant a stockplant area?

(a) Because stockplants in traditional containers grow slower, and are more liable to damage from drought;
(b) To produce many shoots rapidly, in order to harvest plenty of cuttings;
(c) To safeguard clones from the risk of extinction;
(d) To multiply the numbers of plants in new clones;
(e) To reduce the need for travel, especially to distant coppice stumps;
(f) To have stockplants in or near the propagation nursery, so that:
   (1) cuttings can be taken and set on the same day, early in the morning (A 41-44);
   (2) planning and managing the stockplants are more straightforward;
   (3) controlling weeds, shade, soil fertility, watering, etc is easier;
   (4) protecting the stockplants is easier than with coppice stumps (A 21);
   (5) inoculating them with mycorrhizal fungi or nitrogen-fixers is possible (Manual 3).

What sort of trees should I plant?

(a) the species that are currently planted, plus those under trial;
(b) plants growing in containers, or leafless cuttings direct (A 4);
(c) with *increasing* genetic selection (A 10-13) as work advances:
   (1) any seedling plants available;
   (2) seedling plants of the better provenances;
   (3) rooted cuttings from all coppice stumps;
   (4) seedling plants that have scored well in a nursery test;
   (5) rooted cuttings of clones that are promising in a field test;
   (6) rooted cuttings of known superior clones.

Should I plant as many stockplants as possible?

It is generally best to plant most of the available material when starting a stockplant area, to reach a position quickly where many cuttings can be harvested.
However, it is safer to retain a few plants of each clone (or seed-lot) in large containers (A 23).

How should the site be prepared?

(a) If possible, leave some undamaged trees to give moderate overhead shade (A 24);
(b) Cut the undergrowth by hand, and fell unwanted trees with minimum disturbance;
(c) Avoid tearing up or compacting the soil with a bulldozer or grader;
(d) Put in the interplanted shrubs and start planting the stockplants ***soon after clearing***, before the soil has been degraded by exposure to bright sun, wind and heavy rain-drops (A 1).
   [See also Manuals 4 & 5]

How should the planting be laid out?

Here are some general hints:
(a) **size of stockplant area** should be sufficient for a lot of expansion;
(b) **spacing** 1-2 m *between* rows; 0.5-1.5 m *within* each row;
(c) **paths** 2-3 m wide, at intervals;
(d) **road** 5-7 m wide for access;
(e) **clones** clearly separated from each other, e.g by lines of *Leucaena*;
(f) **planting** of clones from one end, leaving room for adding to each.

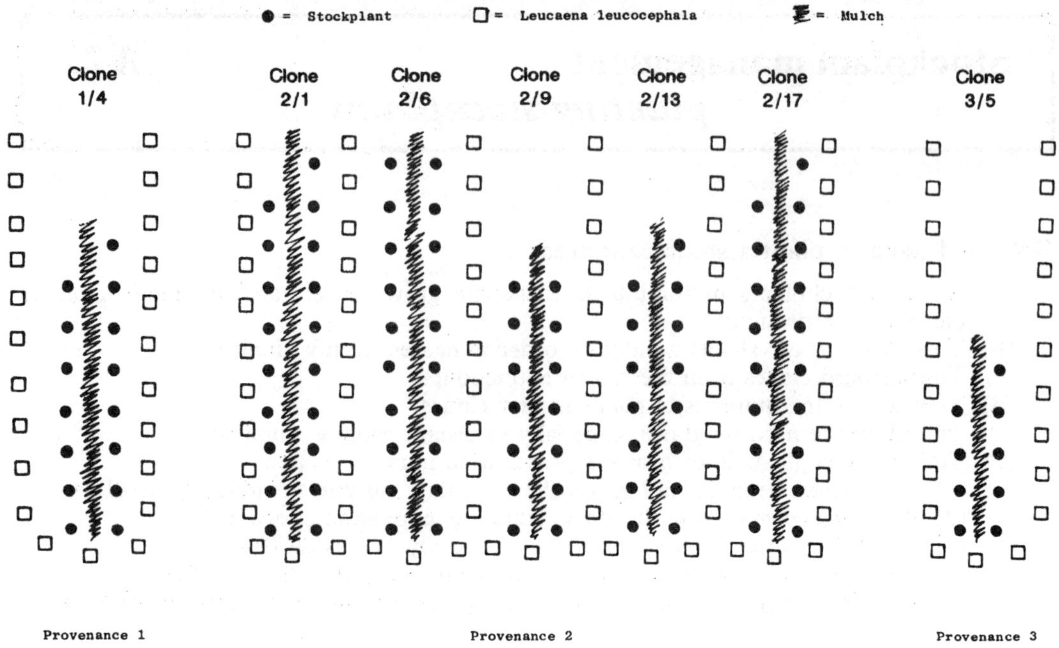

When should the stockplants be planted?

During the normal planting season – also possible at other times of year if regular irrigation is available, or hand watering for small numbers.

Do they need any special conditions?

Multiplication of clones is slow at first, then increasingly rapid. So it is worth spending extra effort to get the early plantings off to a flying start. This can be helped by:

(1) choosing a more fertile site with a gentle slope;
(2) digging larger holes;
(3) interplanting with nitrogen-fixing shade plants (A 24);
(4) mixing in some good soil when planting;
(5) careful planting;
(6) avoiding planting during dry spells;
(7) nursery watering the night before planting;
(8) watering if there is no rain during the first week after planting;
(9) using plenty of mulch (A 27, Manual 5).

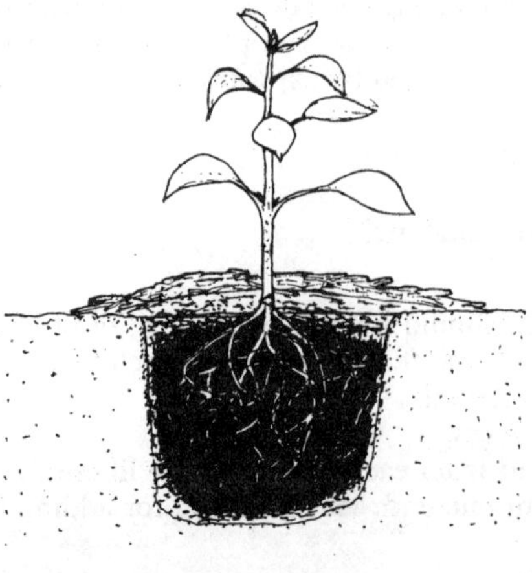

Do planted stockplants need special protection?

Putting up some temporary extra shading, for the first few weeks after planting, will protect your stockplants from stress and drying out before they have grown new roots.
Otherwise, they need the same kind of protection from damage as described for coppice stumps (A 21).

This page has been left blank for field notes.

Stockplant management – *potted stockplants* A 23

Why would I need to keep my stockplants in pots?

Young potted cuttings and seedlings are generally planted out in a stockplant area (A 22), but there may be advantages in keeping some of them at the nursery.

What advantages are there?

(1) **Safety:** clones are less likely to be lost if a few potted plants are held back in the nursery (A 11). It is also easier to keep an eye on diseases and pests (A 52).

(2) **Continuous supply of cuttings:** potted stockplants can be easily watered, fertilised and moved under shade (A 25-27). Cuttings can often then be harvested all the year round, and the type of shoot controlled more easily. At higher altitudes, stockplants could be kept warmer at night in a heated building, if this promotes shoot growth (Manual 3).

(3) **Rapid handling:** because potted stockplants can be moved right into the propagation area, cuttings can be taken, prepared and set without delay (A 42-44). This reduces water loss to a minimum and cuts out the handling damage from using bags.

(4) **Altering the shoot position:** potted stockplants can be laid horizontally or re-potted at a different angle. Changing the orientation of the shoot in relation to the force of gravity modifies *apical dominance*, allowing some of the other buds to sprout and grow out into vertical shoots of the main-stem type.

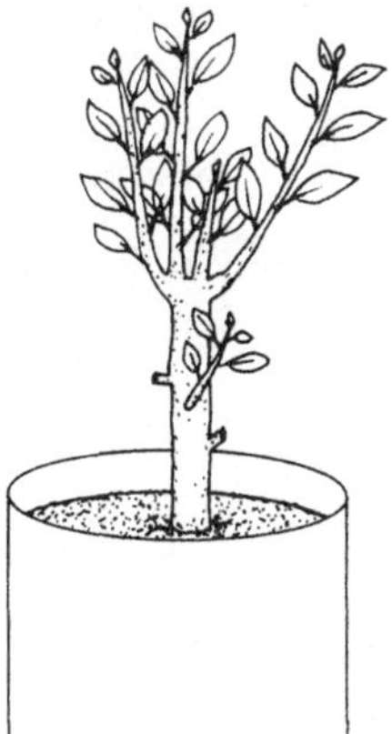

Aren't there some disadvantages too?

The main problems are:

(a) **Slow growth:** the pots greatly restrict growth, so that only a few cuttings are produced.

(b) **Need for regular watering:** unless potted stockplants are watered regularly (A 26), their growth will be further checked, and they could be damaged or even killed.

(c) **Rooting through:** if they are allowed to root through from the pot into the ground, many of the advantages will be lost, and they will be damaged when they are moved later on.

Can anything be done about these disadvantages?

Problems (a) and (b) can be lessened by:

(1) using larger containers (diameter 20-30 cm, or more), and improved potting soils (A 53) to promote better root growth;

(2) using plastic pots that are white outside and black inside, or trying pots made of other materials, to avoid over-heating the root systems;

Problem (c) can be overcome by the simple methods described in A 54.

What kinds of trees are suited to large containers?

Research is needed to find this out, but species that produce many small shoots and leaves (e.g. *Cupressus* and some eucalypts) are more likely to be successful.

What species should not be kept a long time in containers?

Pending research results, avoid those:

(1) with very strong growth and large leaves (e.g. teak);
(2) that are easily damaged by water stress;
(3) which rely strongly on a single tap-root;
(4) are being grown to provide large, leafless cuttings.

What's the best thing to do with an new species?

Plant out most stockplants, but keep some of each clone for trials in various containers.

Stockplant management – *shade for stockplants* A 24

Why do stockplants need shade?

(a) because it protects the soil they are growing in;
(b) because they should grow better and produce more cuttings;
(c) because the cuttings may root better.

How does it do these three things?

(a) by reducing overheating, drying out and washing away of the surface soil, and encouraging the recycling of mineral nutrients (A 1);
(b) by giving soil and lighting conditions more like a gap in the forest or savanna;
(c) by giving the right *amount*, and also *kind* of light to the stockplants.

But won't the shade *slow down* the stockplants' growth?

Only if it is too dense. Moderate shading typically makes shoots elongate ***more*** than they do in full sun.

Why is the kind of light important?

When sunlight has travelled through and past green leaves, parts of it (e.g. the red and the blue) are less well represented. This 'filtered' shade light stimulates longer internodes and more easily rooted shoots.

What kinds of shade can I use?

Two methods using green leaves, one with dying foliage, and one with a man-made product:
Method 1: leave some large trees to provide overhead shade;
Method 2: interplant with smaller shade trees or shrubs;
Method 3: use palm leaf shade;
Method 4: use plastic shade.

Which is best?

Choose which methods are most suitable for each situation – each method has advantages and disadvantages. Advantages include:
Method 1: maintains the whole stockplant area under 'open woodland' conditions; and encourages diversity of mycorrhizal fungi (Manual 3);
Method 2: shade plants can be put exactly where they are needed; and species that are *nitrogen-fixers* can be chosen;
Method 3: very cheap, quick and easy to put up;
Method 4: gives an even shade, and lasts a long time.
Some disadvantages are:
Method 1: uneven shade; dead branches falling can be dangerous and damaging; shade trees sometimes die because of opening up, bulldozer damage, etc;
Method 2: need to be cut back regularly;
Method 3: only lasts a short time;
Method 4: not cheap; needs to be ordered.
For coppice stumps (A 21): methods 1 & 3 are often suited;
For planted stockplants (A 22): methods 2 & 4 are usually best;
For potted stockplants (A 23): use any convenient method.

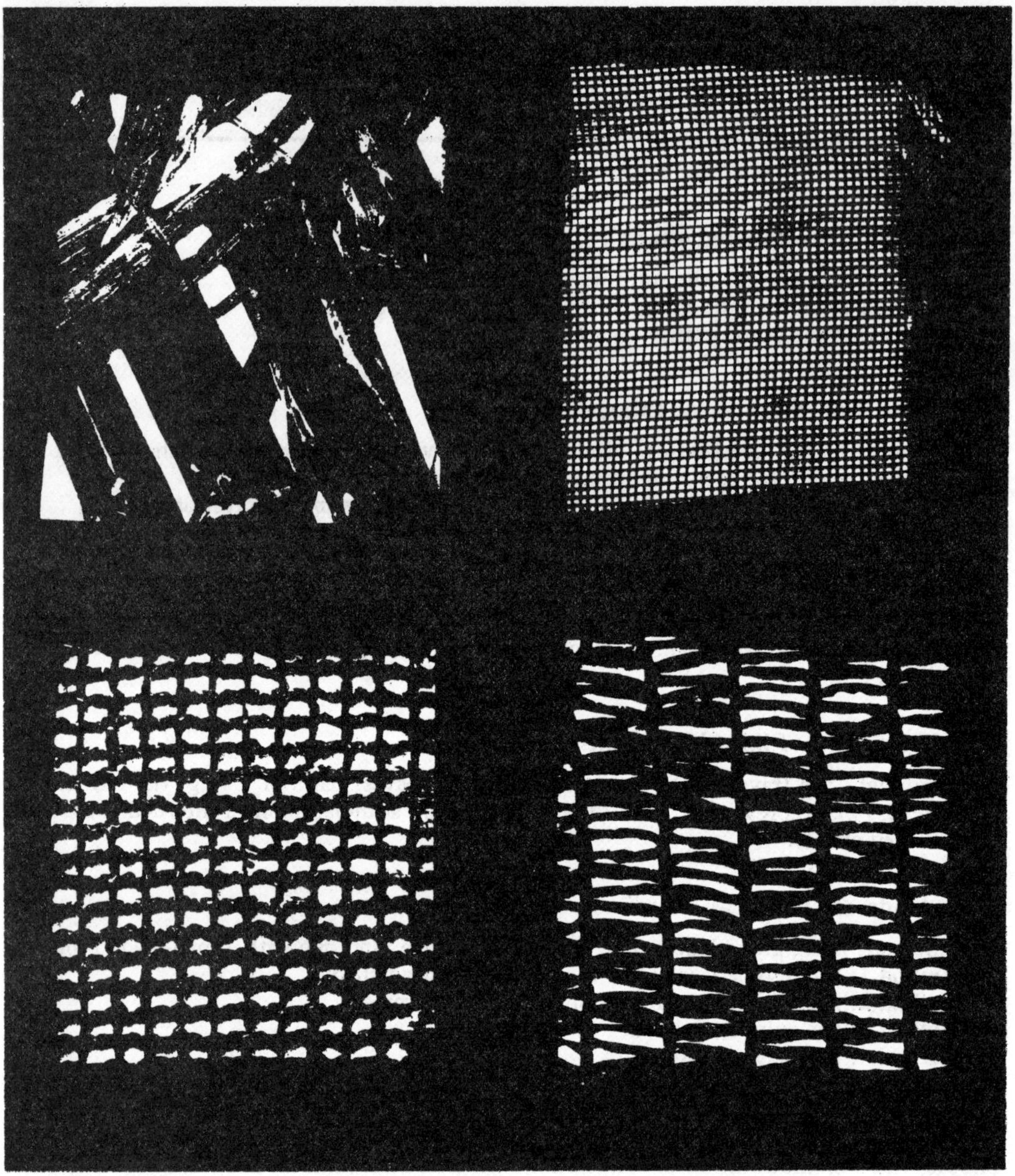

Palm-leaf, and three types of plastic shadecloth.

How do plants fix nitrogen?

Certain kinds of trees, shrubs and food crop plants form ***nodules***, a close association between their roots and micro-organisms. These can transform nitrogen gas from the atmosphere into *soluble compounds* that can be used by the plant (Manual 3). Nearly all are *leguminous plants*, although not all leguminous plants fix nitrogen.

Which kinds are useful in a stockplant area?

Leguminous shrubs, such as *Flemingia rhodocarpa*, which is used in stockplant areas for Robusta coffee, and those that cast a moderate shade, like *Leucaena. Acacia, Delonix, Cassia, Erythrina, Gliricidia, Parkia, Prosopis* and *Sesbania* could also be tried.

What is the best way to plant shade plants?

Close-planted in single lines between 2 rows of stockplants, as this:

(a) gives fairly even shading;
(b) makes it easy to cut them back;
(c) allows one to walk between the stockplants (A 22);
(d) keeps the clones well separated (A 11);
(e) provides nitrogen to each of the stockplants (Manual 3).

What about cutting shade plants back?

Aim for the following:

(1) to cut often, keeping a moderate, even, overhead shade;
(2) to cut less during the dry season, to maintain shade when it is most needed, and when some leaflets or leaves may be shed;
(3) to prune back strong leaders before they get out of reach;
(4) to remove lower branches that get in the way;
(5) to lay the pruned branches on the ground as mulch (Manual 5).

Is plastic shading easy to use?

Yes, the various kinds of *shadecloth* come in rolls up to 4 m wide, and can easily be joined together with a plastic tie (best done on the ground before fixing it up). It is very light in weight, and can be fastened to a simple framework of wood, bamboo, etc, strong enough to stand up to wind and rain. It should then last for several years.

Is temporary shade needed in a stockplant area?

Yes, sometimes, as for example:

(a) to help stockplants or shade plants establish well, immediately after planting;
(b) there is some evidence that shading stockplants for just a few weeks before taking the cuttings could encourage better rooting.

For trials with your species, consider making up some light, portable frames for temporary extra shading.

Potted stockplants can be moved from a shadier to a less shady site.

This page has been left blank for field notes.

Stockplant management A 25
– *stockplant pruning*

Why do stockplants need pruning?

To keep them producing plenty of the right type of shoots (A 6, 20, 21).
Note: *Tying down the shoots of stockplants, or planting them obliquely also affects the number and type of shoots that grow out (A23).*

Doesn't harvesting cuttings do that anyway?

It does part of the job of pruning. However, extra cutting back is usually needed, because:

(1) stockplants tend to re-form strong leading shoots, which:
- (a) are too thick to make good cuttings;
- (b) grow too tall to reach easily from the ground;
- (c) soon lose easy rooting ability;
- (d) suppress the growth of the lower shoots;

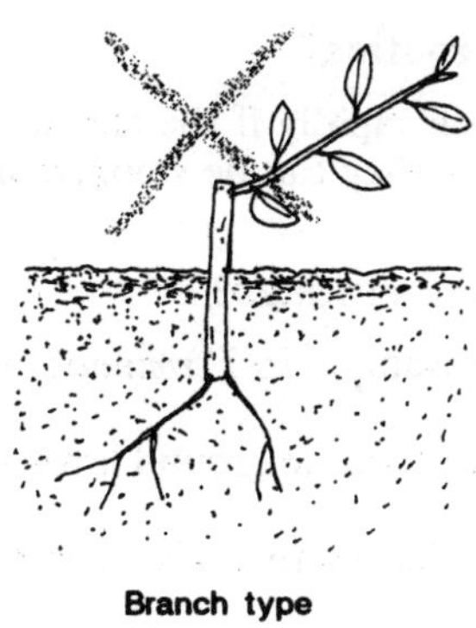

Branch type

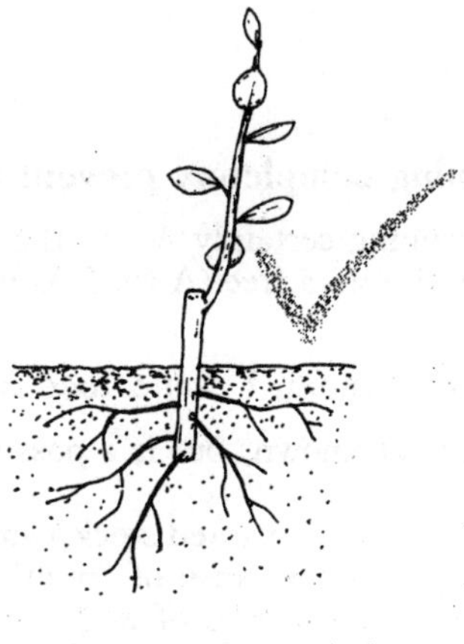

Main stem type

(2) branches need to be checked that are unsuitable for cuttings;
(3) dead parts are better removed.

What should I prune with?

For thinner twigs: a ***sharp*** pair of secateurs or knife;
For thicker stems: a ***sharp*** machete or saw.

When should I prune?

(a) just after harvesting cuttings;
(b) before any shoot turns into a dominant leader;
(c) before new shoots reach about 30-40 cm in height.

How should I do the pruning?

Allow for differences between species, but aim to:

(1) make sloping rather than horizontal cuts, just above a bud;
(2) cut back strong vertical shoots to 2-3 good basal leaves and buds;
(3) remove shoots with a flattened, 'branch-type' growth habit (if the stockplant has few other leaves, just remove their tips and any growing buds);
(4) try removing the tip of the strongest shoot, in order to favour the others.

Can pruning completely prevent the loss of easy rooting?

Regular pruning certainly delays the changes that happen rapidly if the stockplant is allowed to grow back into a tree (A 6). It is not yet clear whether they can be stopped altogether.

Doesn't that mean I could lose my source of clonal cuttings?

This is not yet known, but is a possibility, perhaps for example in *Terminalia*. However, you can:

(a) coppice the planted stockplants, cutting them back near to ground level just before you expect a new flush of growth;
(b) allow a few trees of each clone to grow tall, so that these could later be coppiced, if needed.

Stockplant management A 26
– *watering stockplants*

Why should I need to water stockplants?

There are three main reasons:

(1) to ensure their rapid and successful **establishment**;
(2) to encourage them to grow quickly and produce **more cuttings**;
(3) to enable them to produce cuttings **all the year round**.

Won't watering make the cuttings too 'soft'?

(a) *Not with leafy cuttings*, as they usually root best when the shoots were growing fairly vigorously, and had not yet become very woody;
(b) *For non-leafy cuttings*, it would be sensible to stop watering 2-3 months before you expect to take the cuttings.

But won't the rooted cuttings still be too soft to survive?

No, because they will have been 'weaned' (A 54), grown on and 'hardened' before they are ready for planting out. The 'softness' is just a temporary state, not a permanent feature of the trees.

Will any water do?

(a) **Rainwater:** usually suitable, though watch out for:
(1) supply failing in dry season;
(2) toxic chemicals washed off new roofs into water tubs;
(3) workers using soap in them;
(b) **Mains water:** usually very suitable, even if chlorinated, though it may prove expensive;
(c) **Stream water:** usually suitable, but may run short in dry season;
(d) **River water:** usually suitable, but could contain a lot of iron, aluminium, or industrial pollutants.

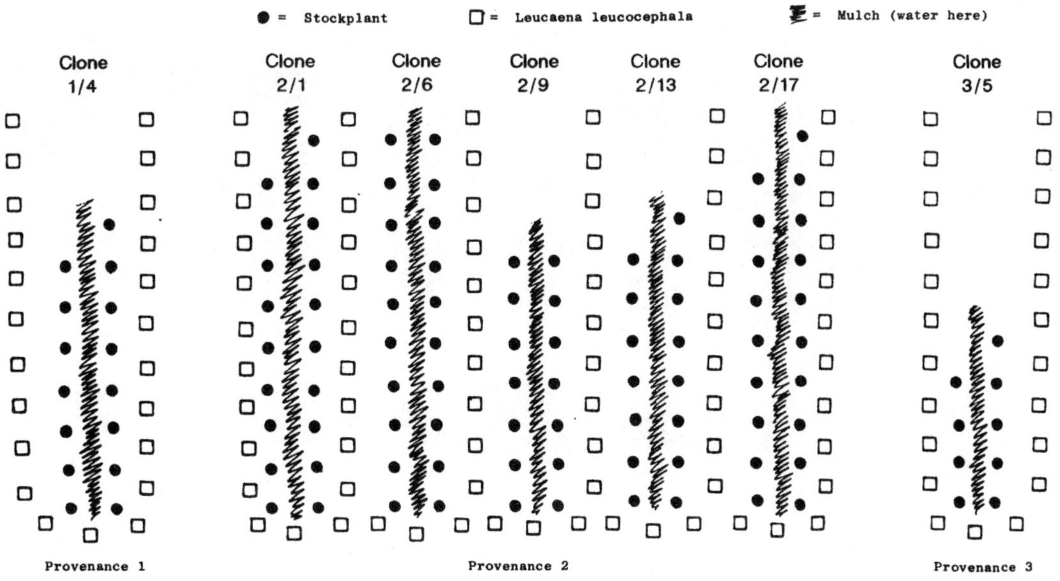

How often should I water in a stockplant area?

(1) *For newly planted trees:* regularly every 1-3 days during the first 2 weeks, unless it rains; (***Note:*** *Mulch should be laid on* ***wet*** *soil)*
(2) *For established stockplants:* plenty of water twice a week is better than a little more frequently.
[NOTE: if you want to stop watering during dry weather, do it gradually!]

What about coppice stumps?

These often grow well throughout the year, probably because they have a very extensive root system, perhaps with mycorrhizas. However, if there is a water supply nearby, an occasional good soaking could be useful for:
(1) stumps that have very few or small shoots;
(2) mulch or fertilisers applied in the dry season (A 27).

How about potted stockplants?

(1) *For stockplants to be planted out:* daily or twice-daily watering, ***including weekends and holidays***, unless it rains (Manual 3).
(2) *For stockplants held permanently in large containers:* work out a watering schedule (e.g. a good soak on Mon/Wed/Fri or Mon/Thurs), which keeps the soil throughout the container moist, but not excessively so.

Stockplant management A 27
– *stockplant nutrition*

Don't trees get all they need from the sun, air and soil?

Yes, these are the sources of the energy and the chemical substances needed for growth of plants. However, mineral nutrients are often in short supply, making growth much slower than it could be.

Does this apply especially to stockplants?

Yes, because:

(a) planted stockplants are usually grown in opened sites, where the supply of mineral nutrients is declining;
(b) rapid growth is needed to produce plenty of shoots that will give good cuttings;
(c) quite a lot of nutrients are being removed by repeated pruning and harvesting of cuttings, and in weeding;
(d) stockplant root systems may be restricted by the removal of much of the shoot system.

What can I do to overcome these problems?

(1) Leave some **shade trees** (A 24);
(2) Interplant with **nitrogen-fixing shade plants** (A 24);
(3) Put plenty of **mulch** around the stockplants (Manual 5);
(4) Apply **fertilisers** to stockplants, including coppice stumps (A 21);
(5) Inoculate stockplants with the appropriate **mycorrhizal fungi** (Manual 3).

Which is the best method?

A combination of techniques, starting with the first.

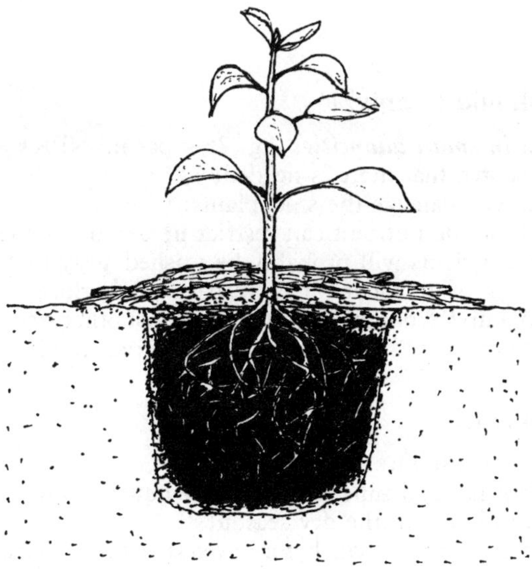

Does that mean that mulch is more important than fertilisers?

Yes, because:

(a) restoring a *woodland environment* comes before trying to compensate for its loss;
(b) mulching lowers the temperature of the surface soil and helps maintain its moisture content, as well as adding nutrients;
(c) mulching adds organic matter to the soil;
(d) mulching laid on wet soil makes the conditions more favourable for root growth, earthworms, mycorrhizas etc and encourages the recycling of nutrients;
(e) using branches from interplanted *nitrogen-fixing* shade plants (A 24) as mulch adds extra nutrients to the stockplant area;
(f) applying fertilisers is generally more expensive than mulching.

So I shouldn't use fertilisers at all?

Yes, they can be a valuable addition to the other techniques, especially if:

(1) the soil is infertile – short of most nutrients;
(2) there is a lack of specific main or trace elements;
(3) material for mulching is in short supply;
(4) research shows that they make the cuttings root better.

What fertiliser should I use?

Try a general purpose fertiliser first. It should contain N, P & K (the main nutrients – nitrogen, phosphorus and potassium) in amounts and proportions that have been selected for agricultural crops (if possible tree crops). Calcium (Ca) and sulphur (S) are usually present in NPK fertilisers; magnesium (Mg) and trace elements (micronutrients) might be required in some cases. Modern slow-release fertilisers may prove well suited to tropical conditions, where heavy rainfall can quickly dissolve and wash away most of the nutrients that have been applied. At present, however, they are more expensive.

What are trace elements?

They are chemicals that are essential for normal plant growth, but are only required in minute quantities. In fact, many of them are toxic to plants at higher doses. They are:

Boron (B)
Copper (Cu)
Iron (Fe)
Manganese (Mn)
Molybdenum (Mo)
Zinc (Zn)

How much fertiliser should I apply?

It is better to apply them ***in small quantities*** (eg. 25 g per m^2 NPK):

(a) until trials have shown that more is needed;
(b) because too much can damage the stockplants;
(c) because too much of one nutrient can restrict uptake of another;
(d) because the extra nutrients will probably be washed away by rain, and wasted;
(e) because high levels of nutrients may restrict mycorrhizas, which are particularly important in taking up P from the soil that would not otherwise be available to the trees;
(f) because large amounts could increase soil acidity (reduce pH).

When should I apply them?

(1) when planting (P fertilisers);
(2) after the stockplant has had time to form some new roots (others);
(3) during the rains, rather than the dry season;
(4) after taking cuttings, rather than before *(unless fertilisers have been shown to promote subsequent rooting)*.

How should fertilisers be applied?

(a) ***little and often***;
(b) P is best applied at the bottom of the planting hole, near to, but not touching the roots;
(c) N and K (and NPK) fertilisers can be applied as a *top-dressing* on the mulch or soil around the stockplant, but not touching its stem or roots;
(d) If the stockplants are being watered (A 26), nutrients can be applied dissolved in the water. Apply frequently at low concentrations;
(e) Trace elements and urea can also be sprayed as a very dilute solution on to the leaves *(foliar feeding)*.

How big a difference does nutrition make?

Except in very fertile, favourable sites, the differences can be very large indeed. In some cases, applying mulch and/or fertilisers could double or triple the production of cuttings.

This page has been left blank for field notes.

Propagating conditions – *introduction* A 30

What conditions do leafy cuttings need?

(1) Very high air humidity;
(2) Moderately low light intensity;
(3) Equable temperatures;
(4) A suitable rooting medium;
(5) Protection from wind, heavy raindrops, pests and diseases.

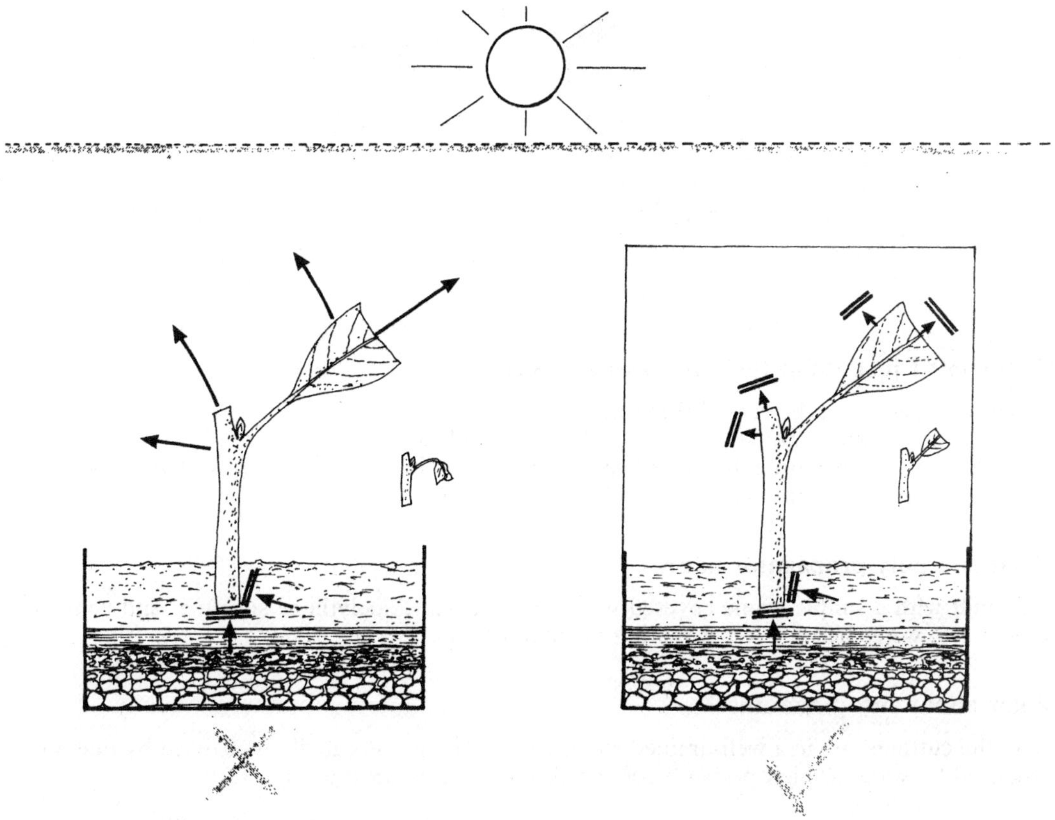

Arrows show water movement; double lines show barriers to it.

Which is the most important?

Lack of any one requirement can damage or kill cuttings. However, ***keeping them always at a very high humidity*** (A 34) is the most important, because:

(a) most unprotected leafy cuttings lose a lot of water within minutes, even in the humid tropics, by *evaporation* from their leaves (and stems);
(b) at first they have no root systems, so water *uptake* is slow;
(c) even when roots have begun to form, they might not absorb enough water to replace what would be lost by evaporation in the open.

How do I keep cuttings from drying up?

(1) By shading the propagation area (A 33);
(2) By trimming the leaves (A 41) before taking the cuttings (A 42); and then putting them straight into a polythene bag;
(3) By taking them when it is cool and humid (A 34);
(4) By rooting them in a poly-propagator, or under automatic mist or 'fogging';
(5) By keeping them moist with a hand-sprayer while preparing (A 43) and setting them (A 44), and whenever the propagator lid is open;
(6) By potting up and *weaning* rooted cuttings carefully (A 53-54);
(7) By checking the conditions regularly (A 50).

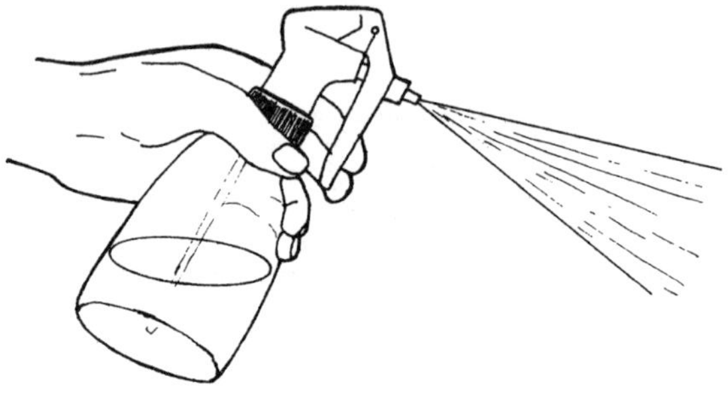

Why can't I just stand the cuttings in water?

A few tree species can be rooted this way, but:

(a) gums, latex or mucilage, or air bubbles, can block entry of water;
(b) the water may quickly go foul, unless aerated as in a 'bubble-bath' system (A 40);
(c) such root systems can be difficult to transfer successfully to soil.

What is a poly-propagator like?

It is built with a simple frame covered with clear (or white) polythene sheeting, and contains a reserve of water below a moist rooting medium (A 31).

What about automatic mist?

Here the cuttings, set in a well-drained rooting medium, are repeatedly moistened by fine mist produced by water at high pressure passing through special mist jets (A 32).

What is 'fogging'?

A greenhouse system in which a powerful fan forces many very fine droplets of water into the air. Most of these evaporate, keeping the humidity high even when the vents are open.

Which is the best method?

Poly-propagators are best, at least to start with, because they:

(a) are cheaper, and easy to build from locally available materials;
(b) do not rely on piped water under pressure;
(c) are simpler to operate, with less risk of breakdown;
(d) can give better results than mist with some species;
(e) can be used either for small or large scale propagation.

When are automatic mist or 'fogging' needed, then?

(1) for species that root better when covered with fine drops of spray;
(2) for research on rooting cuttings at a well-equipped centre;
(3) when found to be appropriate for bulk production of rooted cuttings.

What sort of rooting medium is suitable?

One which allows good aeration, retains moisture (but does not get waterlogged) and encourages good root development.
Mixtures of several locally available materials can be made into good rooting media (see A 35 for more details).

Why should I shade the cuttings?

Because:

too bright sunlight may cause overheating of the cutting, or of the air in the poly-propagator, both of which can dry out or damage the leaves.

But:

too dim light does not allow the cutting to manufacture much sugar by *photosynthesis*, which is important for rooting; and it encourages the development of disease.

What is the best way to shade the cuttings?

This is described in A 33.

I can't control the temperature, can I?

Adequate shading should usually stop the air temperature in a poly-propagator from rising above 28-33°C, and the temperature of the rooting medium above 25-30°C.
At higher altitudes and latitudes, try to choose a *warmer* site, as air temperatures below 16°C, and bed temperatures below 21°C may depress rooting. *(If electricity is available, flexible cables and* thermostats *are available (A 63) to warm the rooting medium to an optimal level – often 25-30°C, while leaving the air cooler.)*

Do I need to measure the conditions?

A thermometer with a range from 10-50°C is useful, and there are dials which give a rough idea of *relative humidity*. When the sky is overcast, a camera at a fixed distance can be used to compare light reflected from a large sheet of white paper. For instance, if the light meter calls for an exposure that is twice as long, then the light level is about 50% of the previous reading.
*(**Note:** Sensors that measure temperature, relative humidity and light accurately may be needed for research.)*

What about protection from wind?

(a) Put up screens made of polythene sheeting, shadecloth, mats, etc, to protect the propagation area, especially the place where cuttings are prepared (A 43);
(b) grow hedges and/or lines of trees to check wind movement, especially to prevent disturbance to automatic mist systems;
(c) secure poly-propagator lids, temporary covers, shading and screens against storms.

What about heavy raindrops?

These can knock over or damage cuttings in mist propagation beds. The overhead shading will reduce the impact of the rain; if necessary build a sloping 'roof' of polythene sheeting to direct water off the beds.
Lids of poly-propagators may be damaged by pools of standing water.

How should I protect against pests and diseases?

This is described in sheet A 52.

What about leafless cuttings?

The rate of drying up is much slower with these, especially if the stems are no longer green, but are protected by a layer of bark.

They don't need a special propagation conditions, and can be rooted in the open ground (A 4).

Propagating conditions – *building a poly-propagator* A 31

How big should poly-propagators be?

They can be built to suit the conditions, materials and space available.
A convenient size is often about 1 m wide x 2-4 m long. The height should be between 0.5 and 1 m, with a *sloping cover.*

What materials are needed to build one?

(a) **Frame:** wood (metal or concrete are alternatives);
(b) **Covering:** a thick grade of clear (or white) polythene sheet;
(c) **Filling/drainage** (below rooting medium): stones, gravel, sand;
(d) **Fixing materials:** nails, office stapler and drawing pins to join and fix polythene sheeting, hinges + screws, clips to secure cover against storms.

How much of them is needed?

To build one wooden poly-propagator 1 m wide, 3 m long and 0.5-1 m high:
(1) **Wood:** 250 x 25 mm – 8 metres length;
50 x 50 mm – 10 metres length;
50 x 25 mm – 32 metres length.
(2) **Polythene sheeting:** 2 m wide – 10 metres length.
(3) **Stones** or broken cement blocks (30-120 mm) – 0.5 m^3;
Gravel (5-10 mm) – 0.25 m^3
Coarse sand – 0.25 m^3.
All the filling/drainage material needs to be thoroughly washed before use.

Won't the wood rot?

Durable, termite-resistant wood should be used, especially for the parts in or resting on the ground. Alternatively, the wood should be treated with a preservative that will not damage the cuttings.
You could attach the polythene sheeting on the ***inside*** of the framework, so that the wood is not permanently wet.

What should I do when the timber frame has been made?

Attach the polythene sheeting, making *doubly overlapping* joins between one sheet and the next. Use a single piece without holes for the whole of the base plus the lower 0.3-0.4 m of the sides. This should be left loose enough so that when it rests on the ground it will hold the filling/drainage without excessive strain.

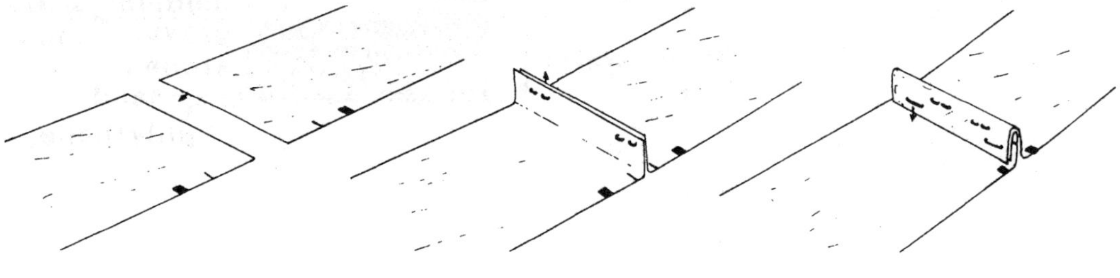

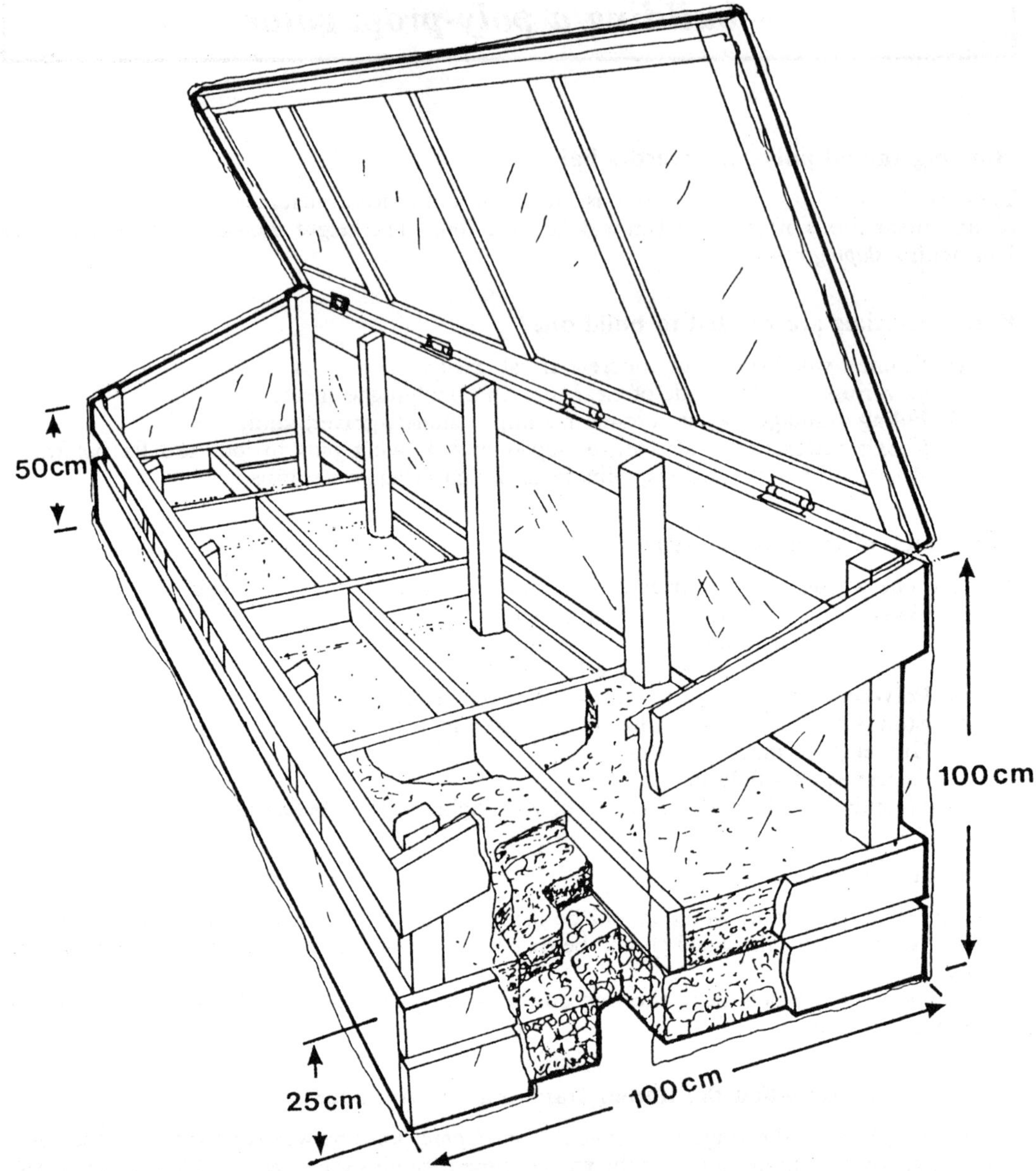
50cm
100cm
25cm
100cm

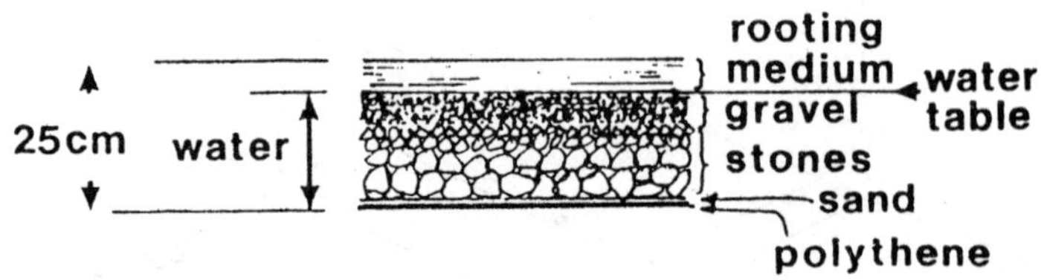
25cm
water
rooting medium
water table
gravel
stones
sand
polythene

Is the poly-propagator now ready to fill?

Not yet. First align its long axis east-west, and then dig four small holes to anchor the 'legs' at the corners of the propagator;
Level out the ground between these holes, and spread sand to protect the polythene sheet from getting pierced or stretched;
Make sure the propagator stands level, and then fix it in place with stones;
Put a short piece of plastic pipe or bamboo vertically in the corner (25-30 cm long and about 5 cm in diameter). This will let you check the correct water level easily, and add water if needed, without soaking the rooting medium.

What do I put in first?

Add ***carefully,*** so as not to damage the polythene sheet:

(1) a thin layer of **sand;**
(2) a thick layer of **stones;**
(3) a thick layer of **gravel;**
(4) a thin layer of **sand** *(adding up to a total of about 15-20 cm).*
Then add **water** until the filling/drainage layer is fully saturated.
(Mark where this level is in the plastic tube.)

What about the rooting medium?

Now add about 10 cm depth of rooting medium on top (see A 35). It should be moist but ***not*** waterlogged, or the cuttings will not thrive.

What about the cover?

This should be set at a fairly steep angle:

(a) to encourage water droplets that *condense* on the underside of the polythene sheet to run off, instead of forming large drips;
(b) to make it easier to work on the low side of the poly-propagator.

The cover can either be attached with several hinges, or can be made in two sections, sliding to and fro. *(In both cases it must close tightly!)* Make a simple catch to prevent the lid blowing off in a storm.

Is a misting system needed?

This is a *non-mist* propagator – no need for automatic misting (A 32);
However, a *clean* hand or knapsack sprayer giving a fine spray of water droplets is required, for use:

(a) while cuttings are being set (A 44), or potted up (A 53);
(b) whenever the propagator is opened (A 34);
(c) if the rooting medium needs to be moistened (A 35).

What about shading?

This is essential, and is explained in sheet A 33.

Does the poly-propagator need any maintenance?

Not a lot:

(1) clean the outside of the polythene sheeting each week with water and a soft cloth, to avoid scratching it.
Clean the inside in the same way, preferably just before putting in a new batch of cuttings;
(2) check the water level each week, and add more if needed, down the plastic pipe;
(NOTE: do not saturate the rooting medium!)
(3) patch any holes in the polythene sheet with a small piece of sticky tape, or polythene sheet glued on.

This page has been left blank for field notes.

Propagating conditions – *mist propagation* A 32

What is mist propagation?

The rooting of cuttings carried out under frequent bursts of a fine water spray.

Is this to stop them drying up?

This is the main reason. Little water is lost by the cuttings because:

(a) the *relative humidity* of the air near the shoots is kept very high by the evaporation of the mist;

(b) the leaves and stems are covered by many small drops of water, which further reduces evaporation from the cutting.

The mist also tends to lower the temperature of the shoot, through:

(c) the evaporation of mist which cools the air;

(d) the small drops of water acting as a screen to the sun's rays;

(e) the water temperature often being lower than the propagator.

What sort of shading is best?

Plastic shadecloth (50-70% shade) at 2-2.5 m above the whole area (A 33).

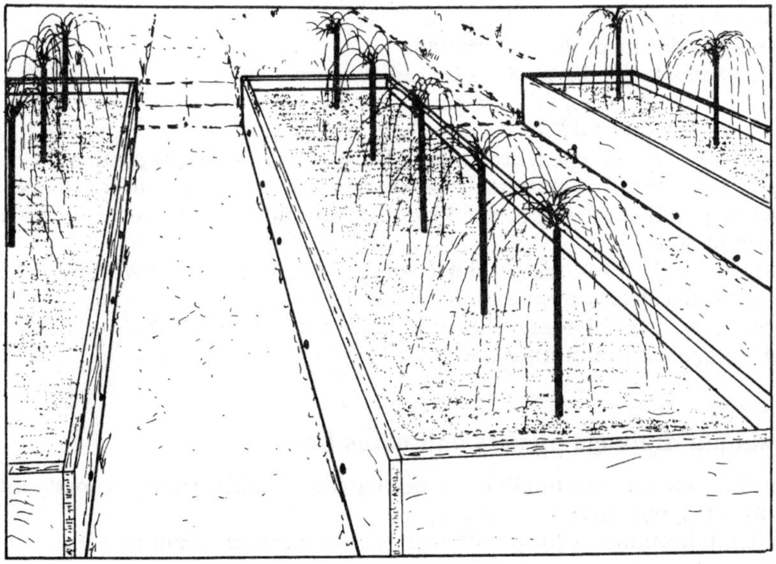

How is the mist produced?

By passing water under pressure through *mist jets,* which are very small, specially made holes. A complete system consists of:

(1) A **water supply** (river, well, reservoir, large tank or piped mains);

(2) An electric or diesel **pump** capable of supplying enough water at a pressure of at least 15 m head, plus a pressure switch and controls;

(3) A **filtration system** which removes particles that would block the mist jets, and can be routinely cleaned;
(4) A **mist controller** (mains- or battery-operated) with a *timer* that allows adjustment of:
(a) the ***length*** of each burst of mist (2-15 secs);
(b) the ***frequency*** of the bursts (every 2-60 mins);
(c) a different regime for day and night;
Alternatively, some controllers have an *electronic sensor* which detects the evaporation of the mist (or whether the sun is shining), and controls the frequency and length of bursts.
(5) A set of **mist jets,** preferably with valves to turn off those that are not needed.

How often should the controller give bursts of mist?

This depends on the climate, the season and the species. A rough guide to start with is:
06.30-18.30 – 10-second bursts every 4 minutes;
18.30-06.30 – 5-second bursts every 45 minutes.

Won't so much water make the cuttings rot?

Too much water can be a problem. It is best to try out different regimes using a common plant species that roots easily, even a weed.
Work out by trial and error how much mist is needed to keep some water droplets on the leaves during the middle of a hot, sunny day; and throughout a night when there is some wind but no rain.

What about when there is heavy rain?

This can also cause problems. These are some of the things you can do:
(1) Make sure that there is very good overall drainage of the site, that can deal with:
(a) heavy rain during thunderstorms;
(b) less heavy but prolonged rainfall;
(c) the water from the mist. *(This can be surprisingly large – to calculate the equivalent daily 'rainfall' caused by the mist, put some cylindrical containers containing a little oil on the bed, and find the average depth in millimetres of water collecting in them during one day. The oil will collect above the water and stop it evaporating.).*
Unless the subsoil is extremely *permeable,* good drainage needs to be achieved by combining sufficient slope, adequate ditches and enough stones & gravel below the rooting medium.
(2) Arrange any overhead plastic shadecloth (A 33) so that the lowest point is over a ditch (or path), not over a propagation bed.
(3) If necessary, lay polythene sheeting under some parts of the overhead shading, as a 'roof' to protect the propagation bed.

How wide should drainage ditches and paths be?

Paths need to be wide enough to allow a wheelbarrow or small trolley to pass; ditches should be adequate to catch and carry away water.
On the other hand, keeping the propagation beds close together means that the relative humidity will tend to be higher.
One possibility is to combine the drainage ditch with the path, by filling the ditch with large stones, then coarse gravel, and a finer gravel on top to make a good, clean path.

How should the propagation beds be constructed?

(a) so that the top of the rooting medium is at least 30 cm above the level of the paths;
(b) construct the sides of the bed out of durable wood or concrete blocks, leaving plenty of space for water to drain out at the bottom;

(c) fill the bed with the same materials as for polythene propagators (A 31), but:
- *without* polythene sheeting;
- with a thicker layer of both stones and of gravel;
- *without* flooding the filler/drainage material with water;
- with a rooting medium that does not easily become waterlogged (A 35).

How wide should the beds be?

Between 1 and 1.25 m, so that one can reach the cuttings in the middle.

Won't the wind disturb the mist?

Yes, this can be a real problem. One answer is to put a vertical screen of polythene sheeting all around the propagation bed, up to 30-40 cm higher than the mist jets. ***(NOTE: do not close off the top!)***
Or you could use an extra line of mist jets upwind of your propagation.
Other suggestions for tackling wind problems are given in A 30.

This page has been left blank for field notes.

Propagating conditions – *shading for propagation* A 33

Won't shading make the trees too soft to survive planting out?

No. As with germinating seeds (Manual 2), shading is used to protect the young plants at a critical stage. Later on, when their root systems have developed, shading is progressively reduced so that the plant becomes 'hardened' to full sun (A 54).

Should the shade cut out direct sunlight completely?

No. It should allow about 15-25% of full light to reach the propagation bed in a poly-propagator, or about 30-50% for mist propagation.

What kind of shade is best for cuttings?

Use local materials such as matting, woven bamboo, palm-leaf or similar leaves. Or you could buy plastic *shadecloth (A 63)*, or metal sheets with some translucent panels. The shading should:

(1) be as ***even*** as possible, avoiding both large *sun-flecks* and areas of dark shade *(having the long axis of the poly-propagators east-west is best);*

(2) ***extend*** at least 2 m further than the propagation beds in each direction, so that unshaded sunlight does not reach the cuttings in the early morning or late afternoon. Alternatively, supplement the overhead shade with mats or pieces of shadecloth hanging vertically, or with palm fronds fixed into the ground.

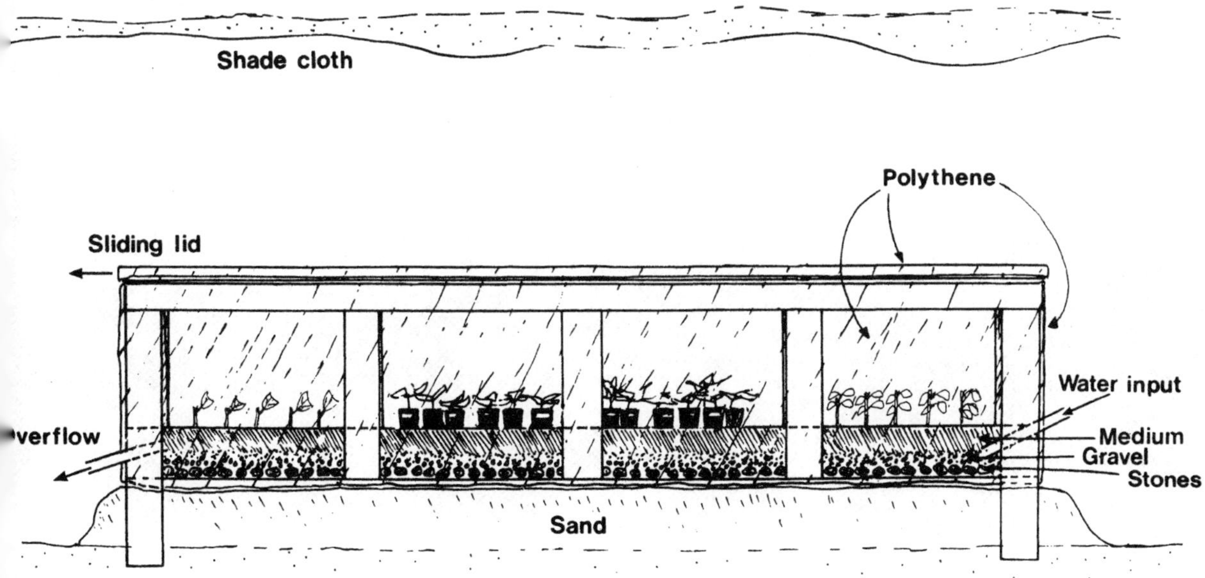

What is the best way to support the shade?

Build a wooden framework at 2-2.5 m, well above head height, over the whole propagation area, including the work area where cuttings are prepared (A 43) and weaned (A 54).
Choose a spacing that fits the sizes of the shading material, and put the uprights at the edges of the propagation beds *(polythene sheeting could protect them from being soaked by mist.)*

Is plastic shadecloth easy to use?

Yes, and it can last for several years if put up and fixed well (see A 24).

How is it fastened?

By tying it at intervals to the framework. It can be easiest to join some of the pieces together before putting them up.

What about leaves and branches falling on the shading?

Leaves can increase the amount of shade cast, and it may be necessary to arrange access to be able to remove them. Damage from twigs and small branches can usually be repaired easily.

When do I need to reduce shading?

(1) *For mist propagation during the rainy season* – less shading may be needed, because it is usually cloudier. Have two layers of shadecloth, and remove one; or take off some of the palm leaves; ***(Watch out for bright sunny spells!)***

(2) *For weaning rooted cuttings that have been potted up* – Make several small, temporary shading frames, and interchange them to increase the amount of light gradually (A 54).

Propagating conditions A 34
– *maintaining high air humidity*

Why is high humidity so important?

Because below about 90% relative humidity cuttings will soon dry up (A 30).

When is it most likely to drop?

(a) in the middle part of the day, especially if the air temperature inside the poly-propagator goes above 30 °C;
(b) immediately the poly-propagator is opened.

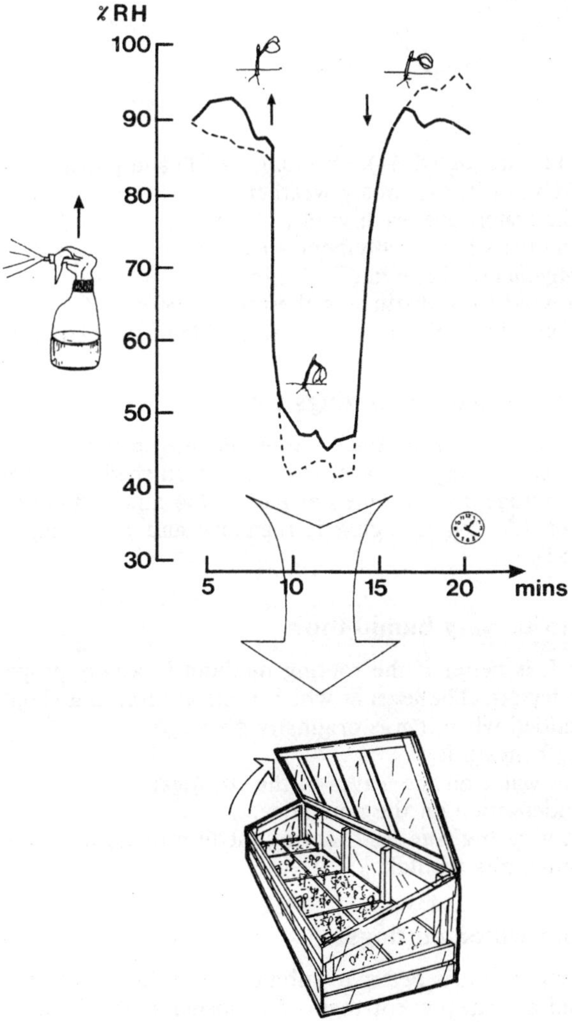

%RH=relative humidity (per cent). Immediately the poly-propagator is opened, the air becomes much drier, even when a hand-sprayer is used - - - -.

How should I keep humidity up?

(1) *in poly-propagators (A 31) by:*
(a) using enough shading, *though not too much* (A 33);
(b) keeping the propagator shut as much as possible;
(c) spraying the cuttings with fine droplets of water weekly, and whenever the propagator is opened *(if necessary each night and morning).*

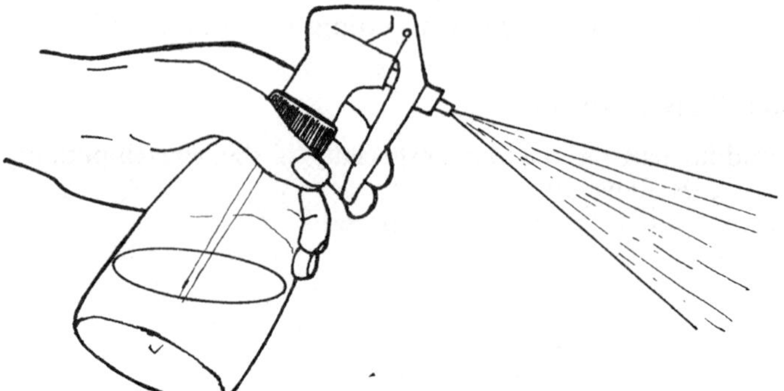

(d) taking (A 42), setting (A 44), checking (A 51) and potting up cuttings (A 53) early or late in the day, or during misty weather;
(e) checking the water level each week;
(f) mending any tears in the polythene sheeting.

(2) *with mist propagation* (A 32) by:
(a) preventing wind from disturbing the mist (A 30);
(b) operating extra lines of mist jets on each side of the beds containing cuttings.

Won't all this moisture make the cuttings rot?

Yes, there is an added risk of fungal attack. However most rotting is not of healthy cuttings, but of those that have been damaged, leaves that have been shed, etc. There are several things that you can do to discourage disease, like not having too much shading (A 33), using a well-aerated rooting medium (A 35), changing it regularly and removing shed leaves and dead cuttings promptly (A 52).

Does the soil need to be very humid too?

No, not at all. In fact it is better if the rooting medium in a poly-propagator is ***not watered*** unless it is starting to dry off. The level at which roots are forming should stay moist from:
(a) the moisture added when it was originally prepared;
(b) the water lying beneath it;
(c) fine spraying of water on the leaves of the cuttings;
(d) drips from condensation on the polythene.

In mist propagation, a very *freely-drained* rooting medium is needed, which does not become waterlogged from the mist plus rainfall.

Can humid air restore wilted cuttings?

They may recover slowly, but it is better to submerge the shoots in water for an hour or two. Afterwards, very humid air can prevent them from losing so much water by evaporation that they wilt again.

Propagating conditions – *the rooting medium* A35

Why do I need a special rooting medium?

Because most cuttings root much better in one than in ordinary soil.

What are the features of a good rooting medium?

(1) **Aeration:** for successful rooting, the base of the cutting has to receive an adequate *oxygen* supply from air spaces in the medium;
(2) **Moisture retention/drainage:** the rooting medium needs to hold moisture, but not be easily waterlogged;
(3) **Mechanical properties:** holes for cuttings should not collapse, cuttings need to be supported, and the roots must be free to penetrate the medium;
(4) **Clean:** Sand, grit and gravel should be washed before use, and other components should not be full of troublesome weeds, animal pests or disease-causing moulds.

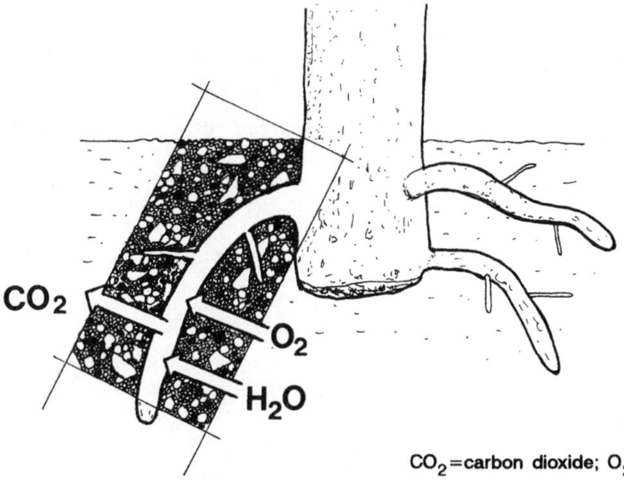

CO_2=carbon dioxide; O_2=oxygen; H_2O=water.

Are rooting media difficult to make up?

Not at all. Many of them are mixtures of common materials, such as sharp sand, grit or fine gravel, old weathered sawdust and coconut palm husks. It is not necessary to buy materials such as peat, vermiculite or perlite, except for special purposes.

Why do I need to mix different components?

Because most of them have weaknesses as well as strengths. For example:
- **grit and fine gravel** are good for aeration and drainage, and can be clean, but are poor for moisture retention and mechanical properties;
- **old weathered sawdust** is good for moisture retention and mechanical properties, but has to be watched for cleanliness.
Note:- most tropical sawdust has to be 'weathered' by leaving it in the open for about a year to break down toxic chemicals.

Can the same media be used for mist propagation?

Free drainage is very important under mist, especially if the beds receive rainfall as well as the considerable amount of water from the intermittent mist. There should be plenty of grit or fine gravel in the mix, therefore.

What mixtures are recommended?

Try 33% grit or fine gravel : 33% sharp sand : 33% old sawdust for poly-propagators ***(moisten the medium while mixing, but do not soak it);***
Try out 50% grit or fine gravel : 50% old sawdust for mist propagation.
The fibre within partly rotted oil-palm stems can be mixed with sharp sand, and old, outer husks from coconuts are greatly prized as a rooting medium.

Will adding fertilisers help the cuttings to root?

No, this might actually hinder rooting, and can encourage small green plants like mosses and *algae* to grow on the surface of the medium.
Although the leaves of cuttings may go a yellowish green because of a shortage of nutrients, the time to replace these is when potting up (A 53) and when feeding nursery plants (Manual 3).

How often should I change the rooting medium?

As soon as there are signs of unexpected rotting, pests or poor rooting.

Propagating conditions – *rooting in containers* A 36

When should I consider rooting in containers?

Problem 1) If the numbers of rooted cuttings you produce is getting larger;
Problem 2) If you find a species with unexpectedly high losses during weaning;
Problem 3) If the root systems in conventional black polythene pots are causing trouble after planting;
Problem 4) If you want to do research on plant handling, growing or planting;
Problem 5) If you would like to transport plants over long distances.

Why is this so?

Problem 1) Because handling and potting up the cuttings could be easier and more successful (A 53);
Problem 2) Because root damage can be minimised with containers;
Problem 3) Because rooting in 'root-trainers' could be the answer;
Problem 4) Because you could reduce the variability of your test plants (A 45);
Problem 5) Because this might offer a 'middle way' between the costs of sending heavy, bulky plants in conventional black polythene pots and the risks of drying out with bare-rooted stock (see Manual 5).

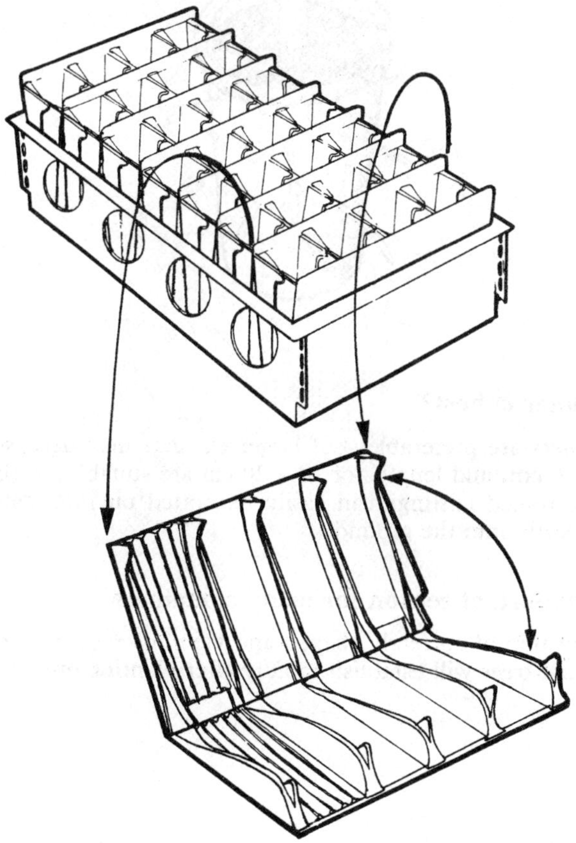

What kinds of containers are suitable?

Any type can be used, including those made of polythene or bamboo but large ones take up too much space in the propagating bed. The best kinds allow free drainage and promote good root systems, such as **root-trainers** and **biodegradable** types. However, these have yet to receive extensive tropical tests, so use them first on a trial basis, filling them with a good rooting medium (A 35).

What is the difference between them?

(a) **Root-trainers** have grooves which make the roots go vertically downwards, and are meant to be kept clear of the bed so that roots do not emerge from the bottom. Each contains 4 – 6 sections, and the whole set can be opened to remove the plants without much damage to their root systems. Around 6 – 10 sets go into a carrying tray for easy handling.

(b) **Biodegradable** pots and tubes are made of material which rots after a period of up to 3 months. Their great advantage is that the rooted cutting is potted up or planted out as it is, *without taking it out of the pot.* Tubes especially need to be supported in trays that will stop the rooting medium falling out of the large hole at the base.

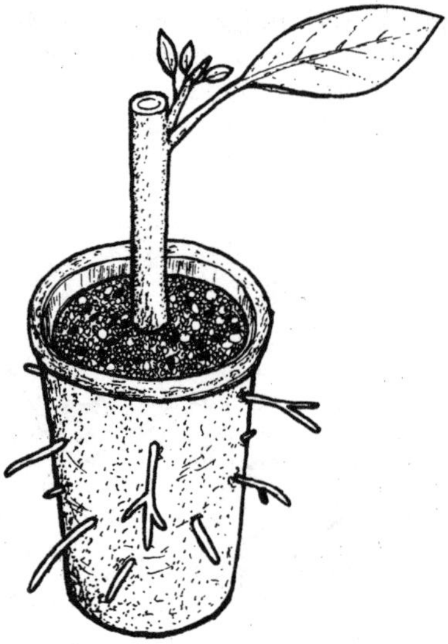

What size of container is best?

Narrow, deep containers are preferable, but larger cuttings may need somewhat wider pots. Top diameters of 2 – 6 cm, and lengths of 10 – 20 cm are suitable, as they don't take up too much space, and the rooted cuttings can easily be potted on into standard polythene pots (A 53), or planted directly into the ground.

What is the most important reason for using containers?

To stimulate the formation of correctly shaped and vigorous root systems which need not be disturbed much, so that trees will establish rapidly after planting out.

Taking the cuttings A 40
– *auxins and rooting*

What are auxins?

They are one of the groups of plant growth regulators (hormones). Some auxins are artificial compounds, but many are produced naturally by plants, especially in young growing parts of the shoot. In nature, they move *always from the tip towards the base of shoots,* but when applied at the base of cuttings there can be some movement in the opposite direction.

What do they regulate?

Auxins control many different aspects of growth and development, as is typical of plant hormones. For instance, they are known to influence the elongation of stems and leaves, the setting and ripening of fruits, and the growth in thickness of trees. ***They also stimulate the formation of new root tips in stem cuttings.***

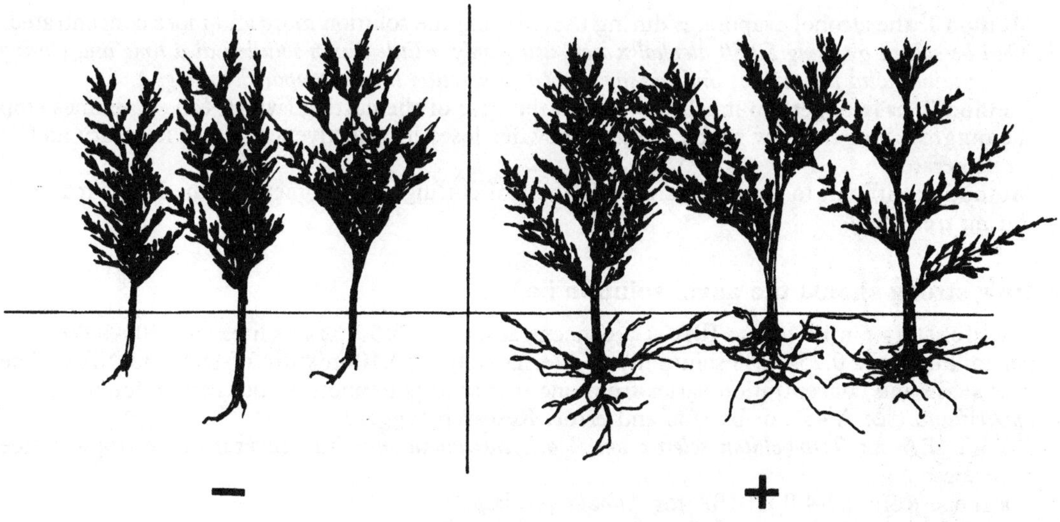

–=no auxin; +=IBA and NAA mixture.

What kinds of auxins are best for rooting cuttings?

Two artificial auxins, known as IBA and NAA, are most commonly used, either singly or mixed together (see A 63 for the chemical names and sources for auxins). Several other auxins are available, but turn to them only if you draw a blank with IBA, NAA and IBA/NAA.

How are they applied to the cutting?

There are three main methods:
Method 1 – Quick dip: dissolve the auxin in alcohol (or another organic solvent), dip the basal few mm of the prepared cuttings for 1-5 secs into the solution and let the alcohol quickly dry off before setting them. ***(Note: If left in alcohol for more than a few seconds, the base will die!)***

Method 2 – Powder (talc): Buy a commercially prepared auxin rooting powder, put the basal few mm of the prepared cuttings into the powder, gently tap to remove excess, and set at once.
Method 3 – Soak: dissolve the auxin in water, and leave the prepared cuttings standing with the basal 2-3 cm in the solution for 4-12 hours before setting them. ***(Note: Many auxins are slow to dissolve in water.)***
With the 'bubble-bath' system, the cuttings remaining with their bases in water, which is kept aerated (as in a fish tank). Small quantities of auxins are dissolved in the water to assist the rooting.

Which is best?

Method 1 has been generally found to give the best results for commercial nurseries producing woody ornamentals in the USA. Advantages include the correct amount of auxin actually entering the cutting, dissolved in the alcohol, and roots forming higher up the stem.
(For research treatment, see A 45.)
Method 2 can be convenient:
(a) for preliminary trials;
(b) if there are no facilities for weighing out auxins; and
(c) before staff and workers have been trained.
Some rooting powders also contain a *fungicide* (A 52, A 63), which helps to discourage disease.
Method 3 is a stand-by in the few species that are intolerant to alcohol.

What are the disadvantages?

Method 1: the alcohol evaporates during use, making the solution more and more concentrated. *(Can be solved by using 50/50 alcohol/water, using only a little auxin solution at a time and closing the container with the thumb, and having a good production line of prepared cuttings.)*
Method 2: easy to get too much powder on the base of the cutting (which can sometimes stop the outgrowth of the new roots), and the powder loses its effectiveness if not kept dry and in a refrigerator.
Method 3: difficult to use with larger numbers of cuttings, and sometimes the leaves may wilt during treatment.

How strong should the auxin solution be?

A useful start can be to use IBA at a concentration of 2.5-5.0 grams/litre (= 2500-5000 parts per million, or a 0.25-0.5% solution), or alternatively a 50:50 mixture of IBA and NAA. The best auxin and concentration varies from one tree species to another, but can be found out by experiment (see A 45), or by trial and error. Research suggests:
2 g/l IBA for *Triplochiton scleroxylon, Vochysia hondurensis* and several other tropical tree species;
a range from 0.5-4.0 g/l IBA for *Albizia guachapele;*
4 g/l for *Cordia alliodora* and
10g/l IBA for *Khaya ivorensis.*
Note: too strong an auxin treatment can discourage the roots from emerging from the stem).

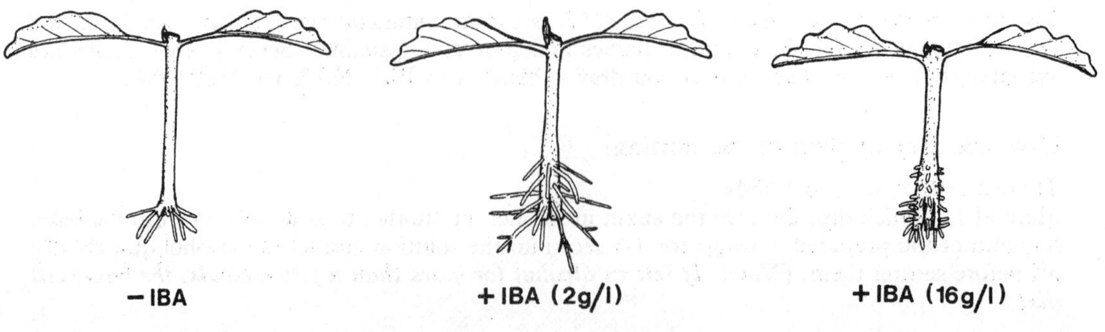

Do all tropical trees respond?

Most do so, but no clear differences in rooting were found in several experiments with *Lovoa trichilioides*.

Supposing I haven't any auxins?

You can still root cuttings of many kinds of tropical tree. Rooting will probably be slower, so concentrate on getting the propagation conditions as favourable as you can (A 30-35). See A63 for sources of auxins.

What do the auxins actually do?

They often:

(a) make cuttings root more rapidly;
(b) stimulate more roots on each cutting;
(c) lengthen the zone in which roots are formed, giving a better root system.

Sometimes they can:

(d) turn a very hard-to-root species into a rootable one;
(e) change the type of root produced, for instance by favouring the rather thicker roots which produce the tree's structural root system.

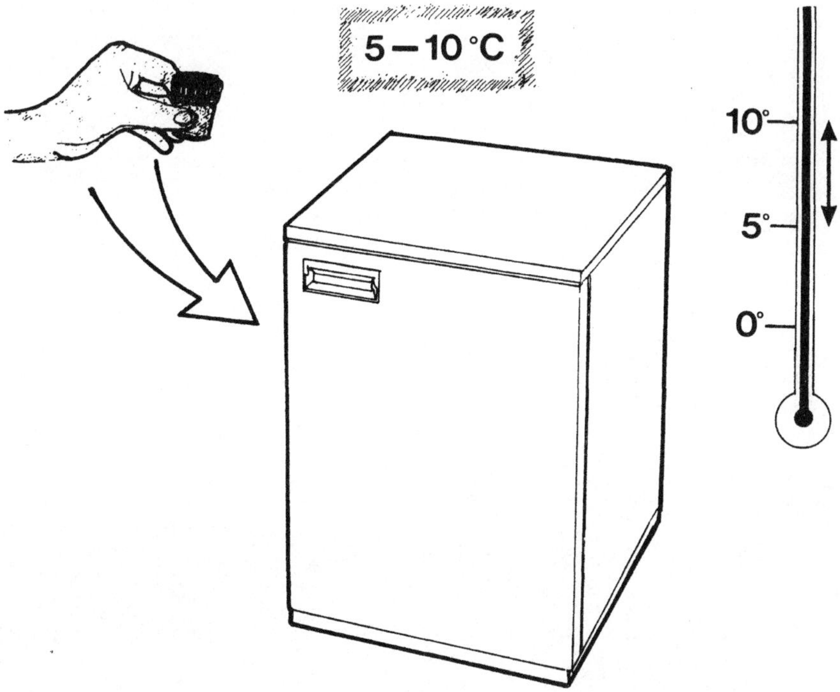

Where should I keep the auxins?

They don't keep well at room temperature in the light, so store all auxins in a refrigerator at 5-10 °C.
Provided they are kept dry as well as cool, pure auxins can be stored for several years, and powders for about a year. *(Note: in the humid tropics, allow the container to warm up for 30 mins before opening it, to avoid getting the contents wet)*. Solutions of auxins keep only for a week or so, even in the refrigerator.

This page has been left blank for field notes.

Taking the cuttings A 41
– *trimming the leaves*

Why should I trim the leaves?

It is important for leafy cuttings of species with leaves longer than about 8-10 cm. Experiments with *Triplochiton* have shown that rooting is better when the leaf-blade is neither too big nor too small.

Why is this so?

Large leaves may lead to more loss of water by the unrooted cutting. On the other hand, small leaves may not produce enough sugars and other substances needed for the cutting to survive. Trimming also makes the cuttings easier to handle, take up less propagation space and be less likely to rot.

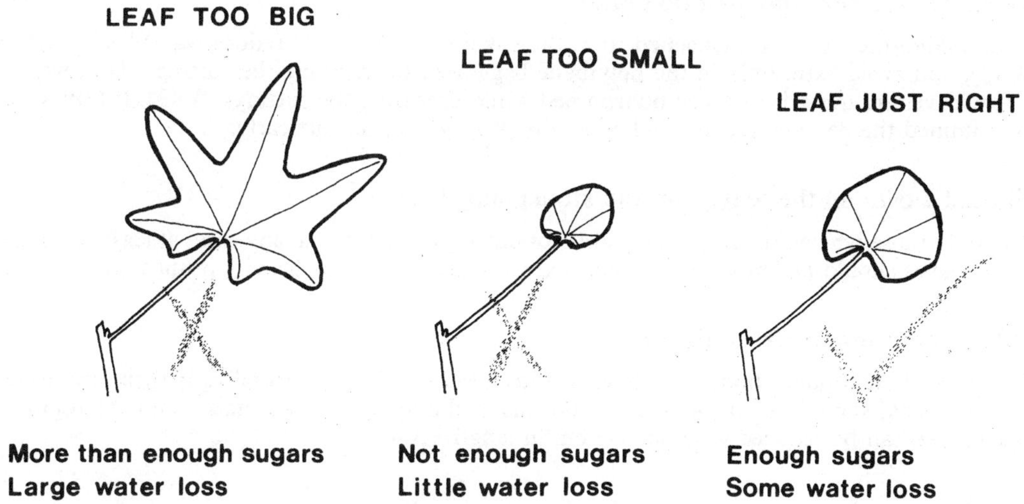

What is the best size?

This varies from one tree species to another. Aim for a total area of about 50 cm^2, except for the mahogany family (Meliaceae). The best size is also influenced by the type of cutting, auxin application used, propagation environment, etc, so it may be worth doing a trial (A 45).

Does the shape matter?

Not very much. With leaves shaped like a hand, it is easy to cut off the 'fingers' of the larger leaves, and with compound leaves to take off some of the leaflets. If insects have eaten part of the leaf, leave more of the rest. For small-leaved species try *multi-node* cuttings (A 42), with more than one leaf above the rooting medium.

What is the best way of trimming the leaves?

With a sharp pair of scissors, avoiding crushing the leaf tissue.

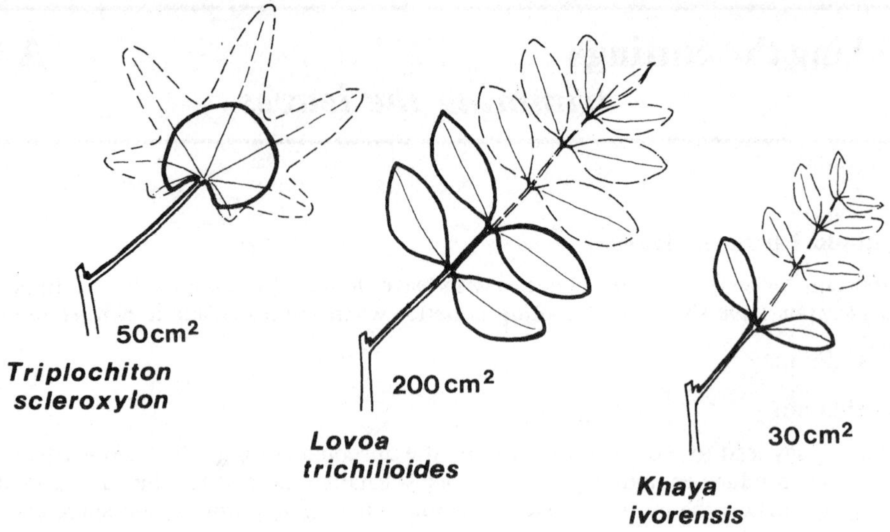

When should the trimming be done?

Trim ***before*** the shoots are detached from the stockplant. This will reduce water-loss (A 34 & A 42), and avoid extra bulk in the polythene bags used for carrying the cuttings. However, if a few leaves are missed, they can be trimmed while preparing the cuttings (A 43). If leaves can be trimmed the day before, this will allow the propagation to start early.

Should I trim all the leaves on the stockplants?

No, only those which are on wood suitable for cuttings (see A 6). In any case, at least two good basal leaves should be left on each shoot, except where it is to be heavily pruned (A 25).

What about non-leafy cuttings?

It is often best to take shoots that have already lost their leaves naturally. If there are leaves present, most should be cut off (best to do this at the top of the leaf-stalk), and the topmost one or two can be reduced to about 1-2 cm in length (A 4).

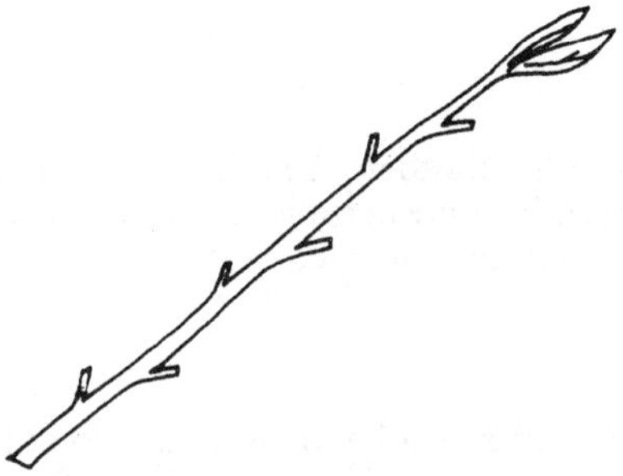

Taking the cuttings — A 42
– *taking leafy shoots*

What needs to be ready before I start?

(1) The **poly-propagator** (or mist propagation area), with shading (A 33), water in the bottom of the propagator (A 31, A 34, A 50) and preparing of the rooting medium (A 35) all done;
(2) A moist, shady **workplace** near the propagation area, with tools and materials ready for preparing the cuttings (A 43) and applying auxins (A 40);
(3) Labels, pencil and **record sheets** ready (A 64, A 66-67); as well as polythene bags, ties and a fine water sprayer (or other supply of clean water);
(4) Except in species with small foliage, the leaves that will left be left attached to each cutting should ***already*** have been **trimmed** to around the optimum size (A 41).
(Note: Get everything ready, so that the cuttings can be kept moist and set in the propagation bed (A 44) without delay.)

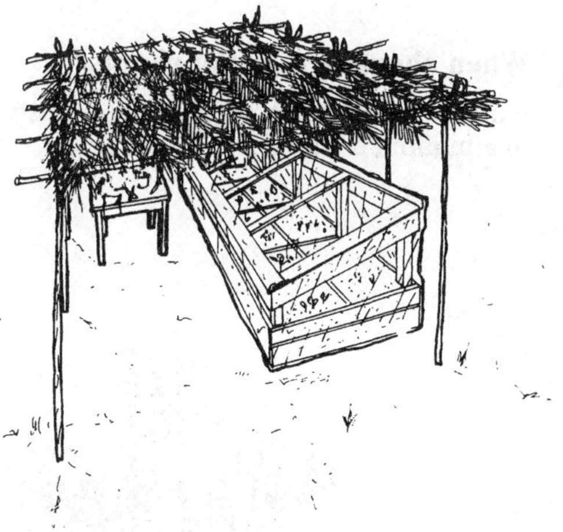

How big should the cuttings be?

As a rough guide, aim to take moderately vigorous shoots that will cut up into cuttings that are 2.5 – 12 cm long, with a stem diameter at the base of about 4 – 8 mm). If possible, avoid thin shoots making little or no new growth; and thick very vigorous twigs, although rapidly growing young coppice shoots are fine before they become too woody. (Non-leafy cuttings are often larger – see A 4).
Also important is the ***number of nodes*** in a cutting, which will depend on the growth habit of the tree species, how the stockplants have been managed and whether you have plenty of material or not.

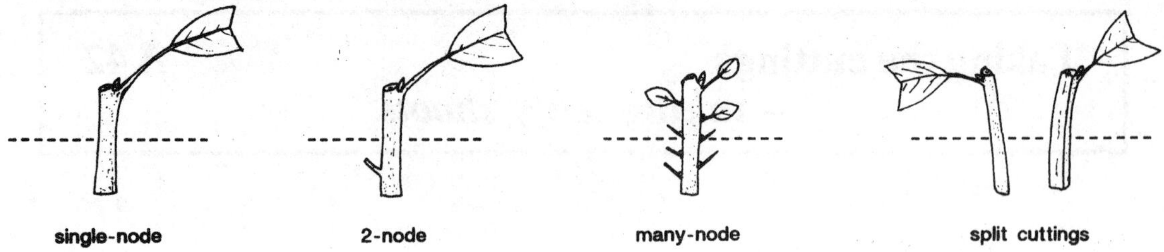

How many nodes should my cuttings have?

(A) **Single-node cuttings** give larger numbers of cuttings from limited amounts of material. Provided that the *internode* between successive leaves or leaf-pairs is more than a minimum of 1.5 cm, cut just above each node that has both a leaf and a bud.

(B) **2 or 3 node cuttings** are suitable for shoots with shorter internodes, those which lack leaves or buds and for species that are harder to root. Cut just above nodes which ***do*** have a leaf + bud, and just below a node, because stems often root better at the nodes.

(C) **Many-node cuttings** are best for species with minute leaves and very short internodes, for example *Casuarina*, *Cupressus*. They are best taken 5 – 15 cm long.

(D) **Split cuttings** are occasionally used where there are two buds at each node. In *robusta* coffee, for example, two cuttings are often made out of one by splitting them down the middle.

When should I take the shoots?

Early in the morning, preferably in misty weather or just after rain. This will keep drying out to a minimum (A 34).

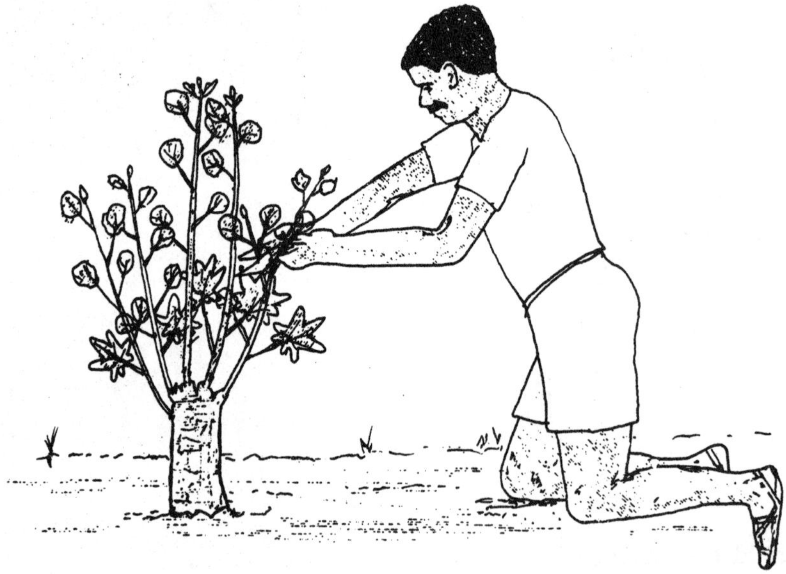

How should I take them?

(a) open a polythene bag, moisten the inside and put in a label with the clone number (A 11), and also write it on one or two leaves with an indelible marker;

(b) cut the shoots from the stockplant with a sharp pair of secateurs or knife, and place them in the bag. ***Keep it closed and in the shade!***

(c) as soon as a bag is completed, spray the shoots lightly, close it with a tie, and attach a second label with the clone number and estimated number of cuttings;

(d) take the bags to the propagation area as quickly as possible. ***(Keep them in the shade and avoid closed vehicles that may overheat in the sun).***
Also avoid piling up or throwing of bags, or the shoots may be damaged. Tests have shown that when trees have been jolted, they often grow poorly. To reduce crushing and bumping of shoots, support the bags in separate boxes or with elastic cords.

What sort of polythene bag is best?

Any kind will do, but the best are of a new type that are white on the outside and black on the inside. The shoots inside are kept much cooler because the white surface reflects heat, while the black one radiates it away. If these are not available, try making double bags yourself by cutting and stapling sheets of black and white polythene.

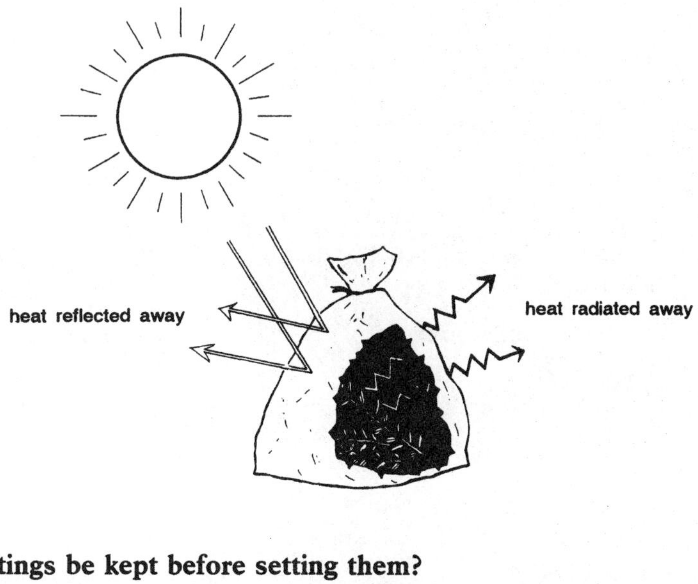

How long can cuttings be kept before setting them?

The sooner you can set the cuttings the better (see A 44). However, avoid opening poly-propagators in the middle of the day, because the humidity inside immediately drops. So plan to set your cuttings before 10 am or after 4 pm. Leafy cuttings of *Triplochiton scleroxylon* will not keep in bags more than 2 nights.

This page has been left blank for field notes.

Taking the cuttings A 43
– *preparing leafy cuttings*

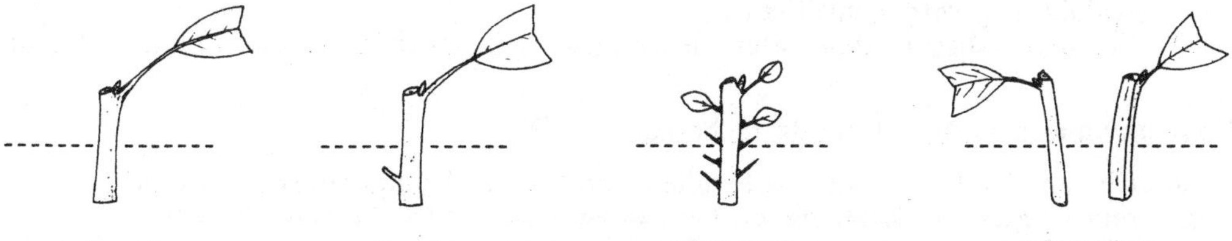

Why shouldn't I set the cuttings immediately?

It *is* very important to avoid delay. However, the **shoots** first have to be **divided up into cuttings,** and there are usually other things to be done before they are set. These include:

(1) **Re-cutting the base;**
(2) **Applying auxin;**
(3) **Removing unwanted material;**
(4) **Treating against pests and diseases.**

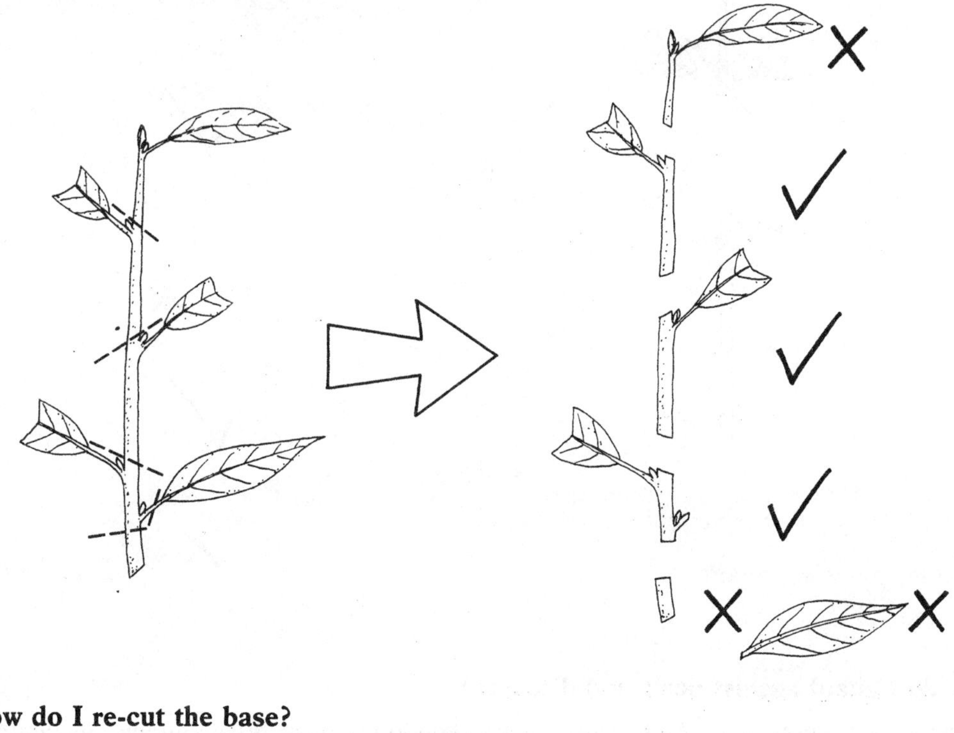

How do I re-cut the base?

Make a new clean-cut base to the cutting, using a very sharp knife or scalpel blade; *in order to*

(a) remove wood that has been split or damaged by the secateurs, which might rot;
(b) shorten cuttings that are too long;
(c) cut the base at a particular angle, e.g at about 90° in *Khaya* – if not, the root system will be one-sided;
(d) cut stems where rooting is most likely (e.g. just below a node);
(e) remove latex or mucilage that may stop the entry of auxins.

What about auxin treatment?

With a fresh-cut base, dip the bottom 5 – 10 mm of the cutting in rooting auxin (A 40) *in order to:*

(a) encourage cuttings to root fast;
(b) increase the percentage of cuttings that root;
(c) increase the number of roots on each cutting;
(i) **rooting powder** – some powder should stick to the cutting, but tap off any excess;
(ii) **auxins dissolved in alcohol** – quickly dip into the solution and dry the alcohol off rapidly (e.g. with a small fan);
(iii) **auxins dissolved in water** – leave standing in water in the shade for several hours.

What unwanted material needs to be removed?

(a) trim any larger leaves missed while trimming (A 41), and cut out any dead parts;
(b) remove leaves at ***lower nodes, but always leave at least one at the top;***
(c) take off side branches that would be below ground;
(d) in trees with two leaves opposite each other, take off *one* of the two buds to produce a plant without a basal fork;
(e) throw out damaged or unsuitable cuttings (A 6).

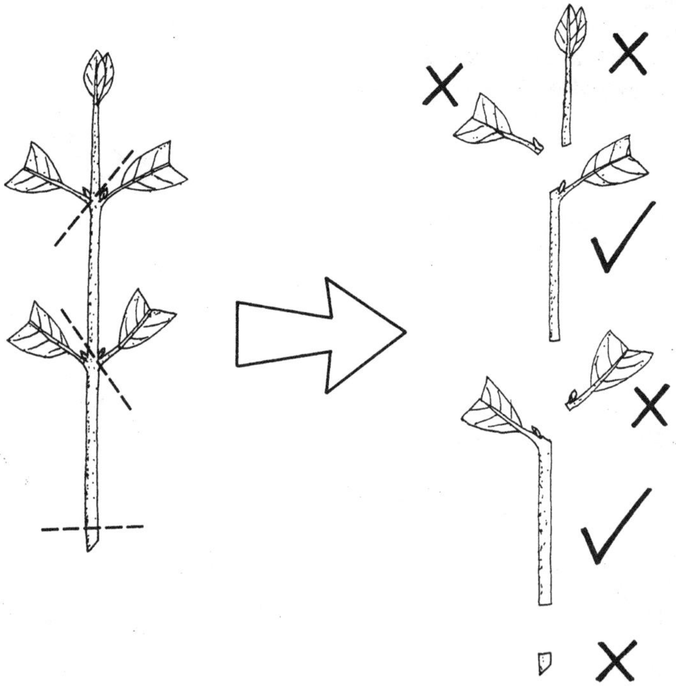

How do I guard against pests and diseases?

If possible, avoid putting unhealthy cuttings into a poly-propagator. Where supplies are limited:

(a) **insects** such as aphids, psyllids, eggs of caterpillars – dip the top of the cutting for a few seconds into a diluted insecticide solution (keeping the base with any auxin powder out of the liquid). Cuttings can then be set in a poly-propagator; but would need to be kept in a moist place for 20 – 30 minutes before receiving mist.
(b) **moulds (fungi)** – treat in a similar way, but with a fungicide solution. Some commercial auxin rooting powders contain 'Captan', a fungicide. (See A 63 and also A 52).
NOTE: insecticides/fungicides may be poisonous – wash your hands before eating!

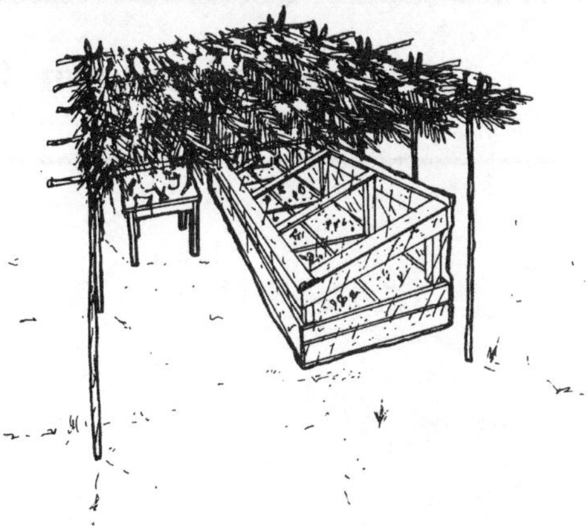

What if I want the cuttings for a rooting experiment?

Cuttings should be as uniform as possible for research (A 45). If you have plenty of shoots, choose a **standard** size and type of cutting. If there is a limited number, **grade** the cuttings into different size classes before allocating equal numbers of each class to the two or more techniques or rooting conditions you wish to compare.

This page has been left blank for field notes.

Taking the cuttings — *setting leafy cuttings* A 44

What should I do first?

(1) Get ready everything listed at the top of A 42;
(2) Check that the rooting medium is moist, **but not too wet** (see A 35);
(3) Trim the leaves (A 41), then take (A 42) and prepare (A 43) the cuttings.

How far apart should the cuttings be?

So that they don't touch each other, but with no wasted space.

How do I put them in?

Make a hole with a small stick (not too deep, and about the same diameter as the cuttings), put in the cutting and make the medium firm around it.

Won't the cuttings dry up while they are being set?

Yes, unless you are careful to:
(a) keep them always in the shade;
(b) leave most cuttings inside a polythene bag while you are working;
(c) use a fine spray of water from time to time (A 34);
(d) open lids of poly-propagators only enough to work, and close them immediately you have finished.

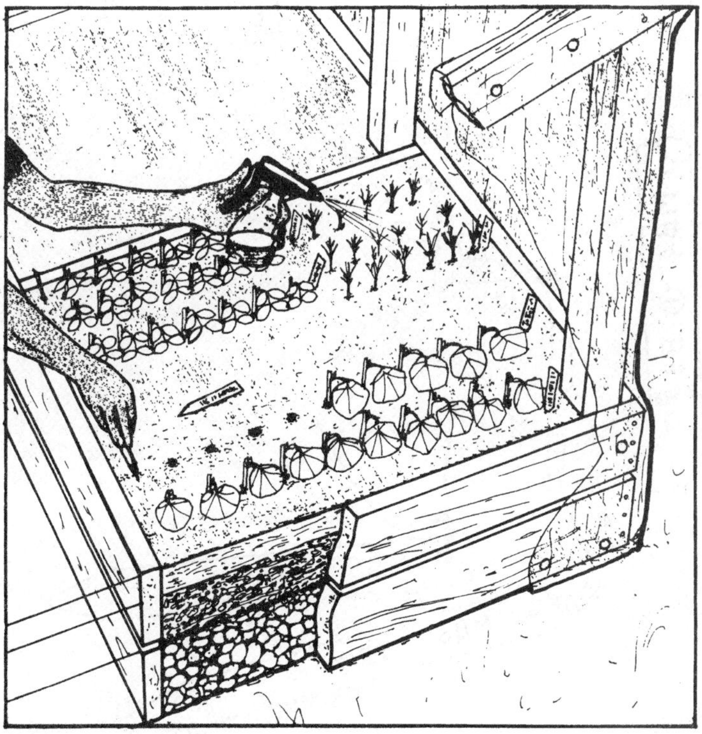

Should the cuttings be in rows?

Rows are much easier, for several reasons:

(1) the separate **clones** are less likely to get mixed up, which could cause long-lasting confusion (A 11);
(2) the new **root systems** need not be disturbed unnecessarily;
(3) there is less **wasted space** if rows of large-leafed cuttings are set with the leaves all pointing the same way;
(4) it is easier to **count** how many cuttings you have set, and check their progress.

What about labels?

At least two labels are needed, one before the first cutting in a batch or clone, one after the last. Write clearly and firmly with a good HB, B or 2B pencil, not a pen.
Essential information includes:

(1) the clone number;
(2) the date of setting;
(3) the number of cuttings set;

In research propagation (A 45), also:

(4) the treatment;
(5) the block; *and*
(6) other variables, such as the origin or type of cutting, node number, etc.

This may require one label per cutting.

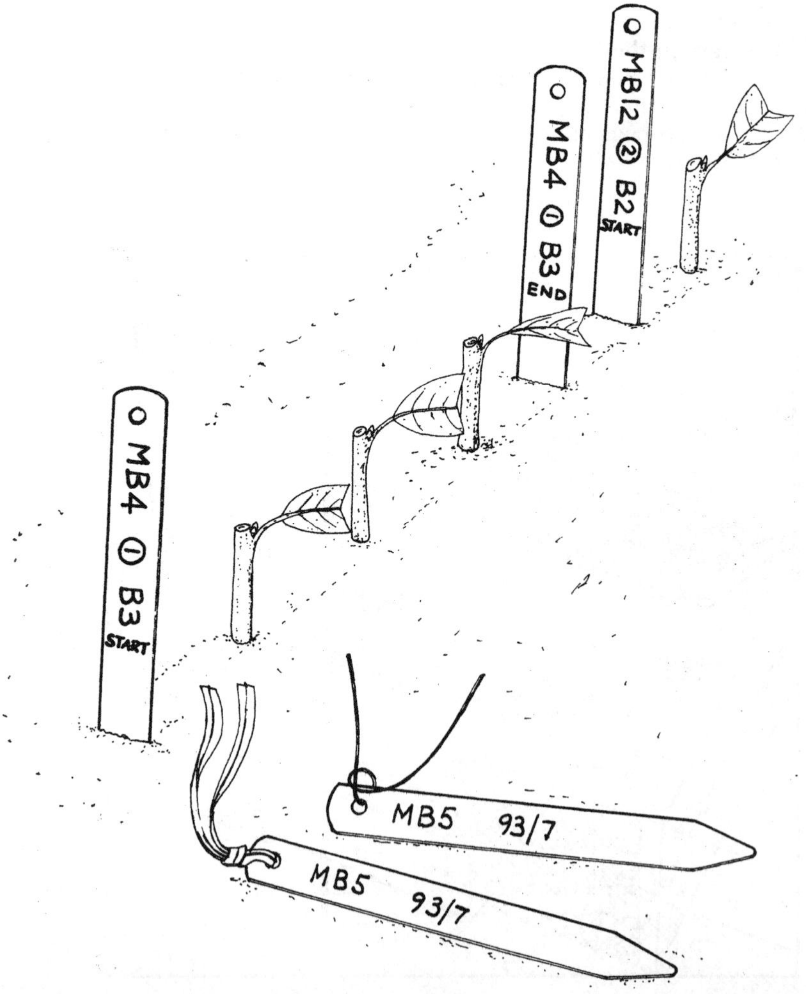

What else is important?

(A) **Cuttings with spreading leaves;**
Watch out that these don't stop water spray or mist reaching cuttings underneath.

(B) **Recording the production of clonal plants:**
Keep records of the origin of each batch of cuttings, and of the clones it contains (see A 66-67).
This will allow calculation of % rooting, % survival, stocktaking and forward planning.
It will also be important to be able to find the stockplants of a promising clone.

(C) **Research on rooting cuttings:**
Keep more detailed records, including a plan showing the layout of the experiment.
Also make periodic assessments and observations (A 45).

How can I speed things up?

(1) Good *planning and preparation* before starting:
(a) Think about how and where people should work, developing a smooth *production line* approach.
(b) Give each person a specific work-rate (eg. 200-350 cuttings per hour for the one just setting the cuttings)
(Note: watch out that the work is done properly!)

(2) Start setting the cuttings at the ***back*** of the propagating bed, and work towards yourself. With wide beds, start at the middle and work towards first one side and then the other.

(3) For setting many small cuttings, make a ***template*** for the holes, using nails driven through a piece of plywood at the correct spacing and depth.

This page has been left blank for field notes.

Taking the cuttings A 45
– *rooting experiments*

What's different about rooting cuttings for research?

It needs greater uniformity, comparability and care at all stages, so that valid comparisons can be made.

Where should I start?

Multiply plenty of cuttings of a few clones, and grow them as well-managed stockplants (A 24-27). In time, you should have a good supply of uniform, clonal shoots.

But surely they will still produce better and poorer cuttings!

Very true. But in any case it is a good plan to **standardise** or **grade** the cuttings for experiments (A 43), whether you are comparing two or more types of cutting; techniques; rooting conditions; or clones.

How and why do I standardise or grade the cuttings?

(1) **Standardising:** If you have plenty of shoots, take more cuttings than are needed for the experiment, and then go through them excluding:
(a) poor cuttings and those with unusual growth habits;
(b) the biggest and the smallest;
because the more uniform the cuttings, the less the variability within the experiment, and the greater the opportunity of obtaining useful results.

(2) **Grading:** If the supply is limited, sort the available cuttings into 2-6 groups, each containing the same number of cuttings of similar size, vigour and growth habit. Then take group 1 and assign equal numbers of cuttings to each treatment, and repeat for the other groups. *This is done because:*
(a) it prevents one treatment being *biased* with many larger or smaller cuttings;
(b) if a note is kept of the group as well as the treatment, it can reduce the *unassigned* variation in the experiment, making it more likely that significant results can emerge. For instance, a treatment could have an effect, but only in the larger cuttings.

But I thought everything had to be randomised!

Not everything. It is the ***individual cuttings within each group*** that need to be randomly allocated to treatments. Grading into groups is meant to be a conscious selection of unavoidable differences, just like choosing where the positions for the blocks should be in your experiment.

What about the propagating conditions?

(1) Try to get the shading as uniform as possible (A 33), and extend it at least 3 m further than the propagation beds in all directions. Alternatively, use vertical palm fronds or hanging 'curtains' of shadecloth all around the bed, to avoid penetration of bright sunlight in the early morning and late afternoon;
(2) Choose those beds (or parts of beds) which are most similar;
(3) Where space permits, avoid using the edges and ends of beds;
(4) Make up fresh rooting medium, and use it for the whole experiment;
(5) Wash the polythene sheeting before setting up the experiment, and wipe it clean as needed.

Example of a simplified layout for rooting experiments

BLOCK I

Clone number	→	MB/12				MB/4				MB/5			
Treatment number	→	4	2	3	1	1	2	4	3	3	1	4	2
node 1	→	*	*	*	*	*	*	*	*	*	*	*	*
node 2	→	*	*	*	*	*	*	*	*	*	*	*	*
node 3	→	*	*	*	*	*	*	*	*	*	*	*	*
node 4	→	*	*	*	*	*	*	*	*	*	*	*	*
node 5	→	*	*	*	*	*	*	*	*	*	*	*	*
node 6	→	*	*	*	*	*	*	*	*	*	*	*	*

Block total = 72 cuttings

BLOCK II

Clone number	→	MB/5				MB/4				MB/12			
Treatment number	→	3	1	2	4	2	1	4	3	4	3	1	2
node 1	→	*	*	*	*	*	*	*	*	*	*	*	*
node 2	→	*	*	*	*	*	*	*	*	*	*	*	*
node 3	→	*	*	*	*	*	*	*	*	*	*	*	*
node 4	→	*	*	*	*	*	*	*	*	*	*	*	*
node 5	→	*	*	*	*	*	*	*	*	*	*	*	*
node 6	→	*	*	*	*	*	*	*	*	*	*	*	*

Block total = 72 cuttings

↓ continue with different randomisation for other Blocks

Can I improve the uniformity of auxin treatment?

Yes, the easiest way is to add one drop (about 50 microlitres) of an alcoholic auxin solution to the re-cut base of each cutting, using a micro-pipette. If you are testing auxin type or amount (A 40), make up solutions at different concentrations, so that you still add one drop to each cutting.
(Note: Alcohol evaporating from the stock solutions will make them more concentrated.)

Do research cuttings need extra day-to-day attention?

Yes. Unrooted leafy cuttings need to be kept in very humid conditions at all times, as they have no roots to take up water. This is especially vital in experiments, because variation in water stress can greatly confuse the results. Uniformity can be helped by the following hints:

(a) spray evenly with fine drops of water; avoid over-watering (A 34);
(b) spray early and late every day, not missing week-ends or holidays and especially during assessments and checks (A 50);
(c) make regular checks on the condition and health of the cuttings, including removal of dead leaves (any dead cuttings removed should be noted in the experimental file);
(d) maintain the shading in good condition;
(e) use a small, flat piece of wood to lift cuttings when they may have rooted; ***don't*** pull them out of the medium.

What layout should I use for the experiment?

(1) A most important point is to have ***plenty of replicates*** in each treatment *(40 is an absolute minimum)*, or real differences won't show up.
(2) For simple, 'look-see' trials, it may be acceptable to ***randomise rows of cuttings*** all of the same type or clone, or receiving the same treatment.
(3) For trials of propagation conditions (e.g. shading, rooting media), a good method is to use a ***single, known clone*** throughout a series of experiments.
(4) For larger experiments comparing clones, the effects of different types or origins of cuttings, or auxin application, etc, it is usually best to use a ***randomised block design***, perhaps with each cutting individually labelled.

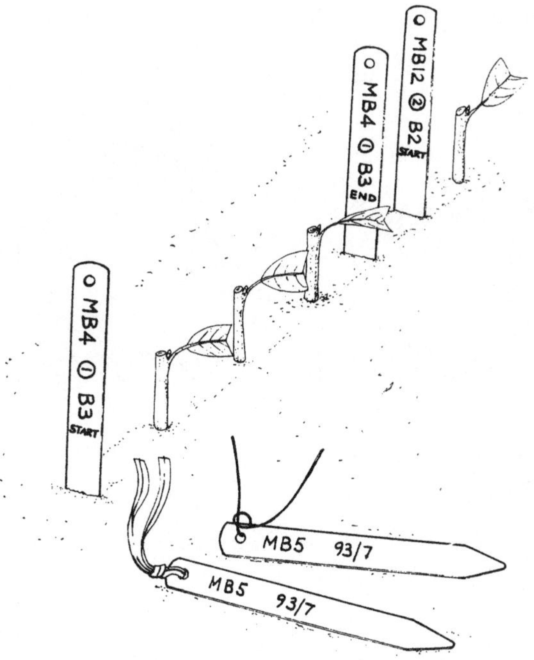

How about labels?

Before taking the shoots, write these clearly and firmly with a good HB pencil, not a pen. *(Even the waterproof types can become indistinct in tropical conditions.)* Labels will be needed for rows or for individual cuttings: abbreviate and condense the writing, but include all the vital information:

(1) the **clone** number;
(2) the **treatment** number;
(3) the **type of cutting** (if this is a variable);
(4) a letter showing the **group** into which the cutting has been graded;
(5) the **block** number, species and **date** of setting can usually be shown on a (separate type of) label;
(6) any **other** relevant information or warning notice.

What about records and assessments?

(A) Records:

(1) Without full notes on the origin of the cuttings and the dates and details of the treatments and propagation conditions used (A 64, A 66 – 68), much of the value of your experiment will be lost.
(Note:- Draw a map showing the layout of the cuttings in the propogation bed, and keep several copies safe in the office.)
(2) Measure the propagation conditions carefully (A 30, A 68), and find out how much they vary:
(a) from place to place;
(b) at different times of day.

(B) **Assessments:**

(1) Record any changes in appearance of the cuttings, with the date.
(2) Carefully lift the cuttings with a small, flat piece of wood every 1 – 2 weeks, and record whether each is rooted, unrooted or dead; and the number of roots (A 65). ***(Do not let them dry up!)***
Pot up well-rooted cuttings, and replace the others in the bed.

What else is important?

A personal input into your experiment – plus the ability to look for explanations for the results other than those you were testing for.

Care of cuttings – *checking the conditions* A 50

What do I need to check?

The most important thing is to see whether suitable propagation conditions are still being maintained.

What needs to be checked regularly?

Points needing to be ***checked once or twice every day:***

(A) **Moisture:**

(1) *in poly-propagators (A 34, 51):*

(a) Unless they are still wet, cover the cuttings with a fine spray of water droplets in late afternoon and early morning especially when the weather is hot and dry. In cool, moist weather, and with well constructed poly-propagators, spray only when the leaves are becoming dry.

(b) top up the water level under the rooting medium, if needed (A 31);

(c) check that the rooting medium is still moist, ***but don't soak it!***

(2) *in mist propagation (A 32):*

(a) check that the mist controller is working properly;

(b) clean any mist jets that are giving uneven or poor mist bursts;

(c) see whether the rooting medium is much too wet. If the shoots are also very wet, reduce the length or frequency of bursts a little.

(d) check for leaves stopping mist from reaching some of the cuttings.

(3) *for potted plants (A 53):*

(a) check whether the soil is still moist, and water accordingly;

(b) look out for any signs of wilting (A 51).

(B) **Temperature:** *in poly-propagators (A 31):*

(a) check that the air temperature is not getting too high in the middle part of the day (A 30);

(b) do the same for the temperature of the rooting medium.

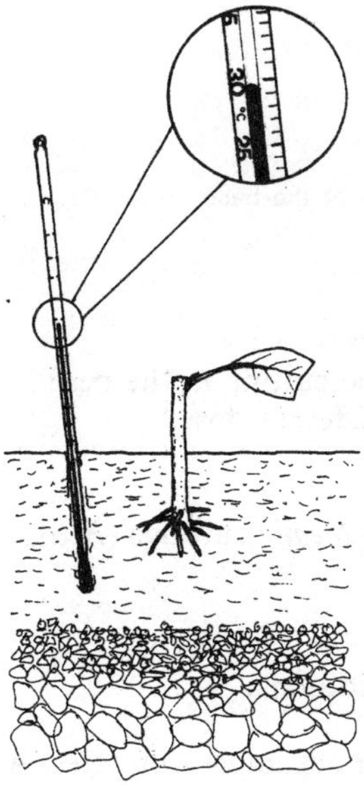

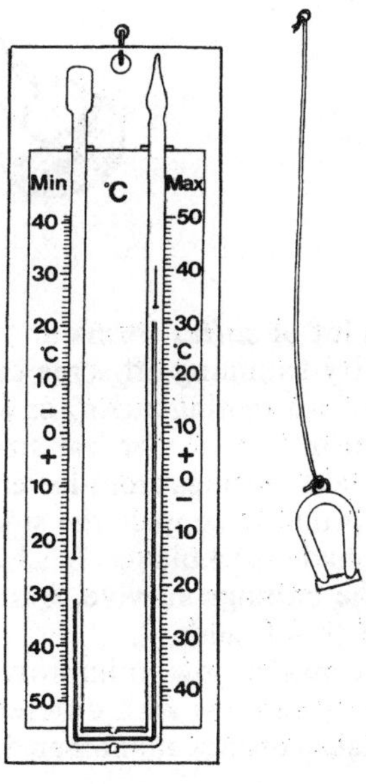

What things can be checked less often?

Make regular ***weekly checks*** on:
(C) **Shading** – that it still covers adequately the propagation beds (A 33), newly potted cuttings, and those being weaned.
(D) **Diseases and pests** – look out for any signs of **rotting,** other than shed leaves or damaged cuttings; and **insect pests** (A 52).

Do the cuttings themselves need checking?

(a) occasionally dig up a few cuttings carefully, to check for rooting.
NOTE: it is best not to disturb the batch unnecessarily;
(b) looking at the cuttings can tell you a lot about whether they are thriving or not.

What can go wrong with them?

(1) **if cuttings shed their leaves** – this could mean that they:
(a) had been too wet in the polythene bags;
(b) were overheated in transit, or damaged by *chilling injury* if too near to ice.
(c) were kept too long before setting;
(d) had been subjected to bright sunlight, drying up or overheating during collection or propagation;
(e) had too little light during propagation;
(f) older, thicker shoots were used, with leaves that would soon have fallen anyway.

(2) **if the base of the cutting rots** –
(a) the rooting medium is (or was) too wet (A 35);
(b) the rooting medium is not well aerated – add more fine gravel, grit or coarse sand;
(c) it is infected with harmful fungi and needs treating with a fungicide (A 52) or changing;
(d) the stem was crushed when the cuttings were prepared – use a **sharp** blade;
(e) the alcohol was not evaporated off quickly.

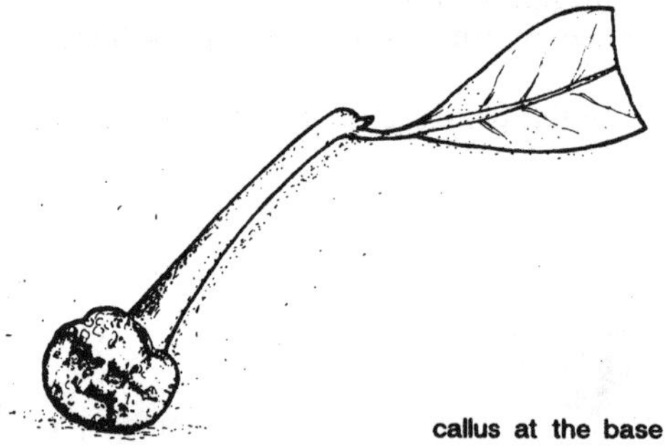

(3) **if a lot of *callus* forms at the base, but no roots** –
(a) try trimming off some callus and replacing the cutting in the bed;
(b) use a rooting auxin, or another auxin, or a different dose;
(c) trim to a smaller leaf size (A 41);
(d) take cuttings from lower on the stockplant;
(e) prune or coppice the stockplant to get more *juvenile* cuttings (A 6);
(f) shade stockplants (A 33).

(4) **if the cuttings survive, but don't root** –
as 3 (b – f) and:
(g) consider how to improve the propagation conditions;
(h) try cuttings with different leaf areas (A 41, 45);
(i) take cuttings at another time or growth stage, for example softer (less woody) shoots.

(5) **if the cuttings go a pale yellowish green colour –**
   (a) check that the shading is not too great (A 33);
   (b) look for signs of pests or diseases (A 52);
   (c) improve nutrition of stockplants (A 27);
   (Note that it is normal for leaves to become a bit paler, as they take up little or no mineral nutrients while they are unrooted.)

What else do I need to look out for?

The most difficult things to guard against are occasional, unexpected problems (A 2). Here are some examples – try and think of others; and what you could do about them:

(1) breakdown of watering over holiday periods, and through illness;
(2) damage from unusual storms;
(3) robbery or vandalism;
(4) flooding from rise in river level, blocked drains, broken water pipe, tap left on or failure of pump controls;
(5) fire damage to wooden framework, shading or buildings;
(6) damage to plants or to installations by domestic or wild animals;
(7) sudden change in weather conditions.

How can I improve standards of plant handling?

(1) By explaining techniques carefully;
(2) By checking the quality of work frequently;
(3) By arranging work schedules with a bonus for good results, rather than a premium for numbers of cuttings set (A 44).

This page has been left blank for field notes.

Care of cutting A 51

– *watering cuttings*

Do cuttings need a lot of water?

Unrooted leafy cuttings ***must*** have high air humidity all the time (A 30, 34), but they will not thrive if the rooting medium is too wet (A 35).

How about when they start rooting?

At this point cuttings can absorb more water from the rooting medium. However, because they are still in a humid environment, the amounts taken up will be small. So you are unlikely to find much drying out of the rooting medium during the regular checks (A 50).

When do cuttings need more water?

After they have been potted up (A 53), particularly while they are being weaned and hardened (A 54).
(Note: during all these stages they need very careful watering.)

How should I water them?

Follow these guide-lines during the critical first 3-6 weeks after potting:

- (a) check ***each day*** whether the soil is beginning to dry up;
- (b) if it is, water the pot well; if not, leave it till next time;
- (c) only water the plants that need it;
- (d) don't 'drown' them with excessive amounts of water; but also avoid adding just a little each time, because this may not reach all the soil in the pot;
- (e) make sure that the water does not run off the leaves and miss the soil;
- (f) if you see any signs of wilting, and the soil is still moist, ***don't water!*** Immediately cover such plants with extra shade, move them to a more humid place or even put them back inside a poly-propagator.
  The problem is not a lack of soil moisture, but rather that the new plant is losing more water from its leaves than it can replace from its still limited root system. *(Check that the roots are not being eaten!)*

Won't plenty of water make the plants too soft?

No, they need extra care until they have had time to grow a sizeable root system. Once they are well established plants, they can be 'hardened' (A 54) in the same way as planting stock raised from seed.

What sort of water is suitable?

The water can come from any source (A 26), provided that it is:

- (1) clean, and without harmful substances in solution; *and*
- (2) at around the same temperature as the new root systems.

Why does the temperature matter?

Young roots are quite delicate, and could easily be damaged by a sudden change of temperature, either hotter or cooler. This is particularly important when the trees have been inoculated with *fungi* to form mycorrhizas (Manual 3).

What can I do about this?

(a) Water early in the day, or late in the afternoon;
(b) Watch out for shallow pools and slowly moving streams that may be hot; and springs, wells and rapidly moving mountain rivers that may be cool.

How can I get water at the same temperature as the roots?

Store it in ***clean*** 200-litre oil-drums in the propagation area. These should be kept under shade, to stop the water heating up. Fill them immediately ***after*** finishing watering, so that the water will have time to cool down or warm up before the next time.

Care of cuttings – *diseases and pests* A52

How can I protect cuttings from disease?

The basic answer is to have good:

(1) stockplant management and production of cuttings (A 20-27);
(2) propagation conditions (A 30-35); *and*
(3) handling (A 41-44) and care (A 50, 51) of cuttings.

But surely diseases could still attack them!

Yes, especially because high humidity tends to favour the spread of disease. However, it is much less likely when cuttings are rooting quickly and well.

How can I tell if disease is starting up?

(a) by regular checks on the cuttings (A 50);
(b) by looking out for discolouration or blackening of ***attached*** leaves or stems – *note that it is normal for shed leaves to go black;*
(c) by occasionally digging out a few cuttings to look at carefully.

What causes disease?

(1) **Fungi** – most kinds of fungus are beneficial (Manual 3), but a few attack weakly or even vigorous plants (for example, root rots and damping-off);
(2) **Bacteria** – again, most sorts of bacterium are harmless or useful, but a few cause diseases such as *cankers* and *wilts;*

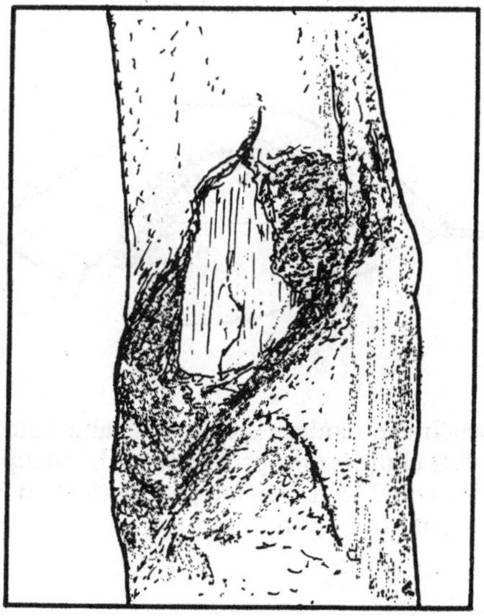

Canker

(3) **Viruses** – these can cause discoloration and distortion of leaves, as in the common leaf mosaic disease of cassava (manioc).

How do they spread?

(1) **Fungi** – by *spores*, which are present in large numbers in the air, water, soil and on (or occasionally within) seeds; *and*
– by *hyphae*, filaments growing in soil and rotting plant material;
(2) **Bacteria** – large numbers of these occur in air, water and soil;
(3) **Viruses** – these are generally transferred to a plant by plant-sucking *bugs*, such as *aphids*, or sometimes by tools. If a virus has attacked a stockplant, it may be present in any of the cuttings taken from it, even if they do not yet show symptoms.

What should I do if I find disease?

Choose the appropriate steps from the following suggestions:

(a) remove all dead or dying cuttings and shed leaves promptly;
(b) if the disease appears to be localised, remove that part of the rooting medium which may have become contaminated;
(c) if it is spreading, spray all the cuttings in that bed with a *fungicide* like ***benomyl*** or ***sulphur*** (A 63);
(d) before setting the next batch of cuttings, use a *drench* of benomyl, or put in fresh rooting medium;
(e) if disease causes persistent trouble, dip new cuttings briefly in benomyl before setting, or use a rooting auxin containing ***captan;***
(f) treat diseased stockplants, and dig out and burn any that remain suspect (if you have some healthy ones of the same clone).
(g) avoid contaminating potting mixtures (A 53), and mulch (Manual 5);
(h) separate potted plants that are affected from healthy ones; potted plants that are affected from healthy ones;
(i) try taking cuttings soon after the shoots have elongated.

What about insect pests?

There are several different kinds, including:

(1) **Plant-sucking bugs,** such as aphids, psyllids, thrips and scale insects, which can weaken the cuttings and/or transfer viruses;
(2) **Leaf-eating insects,** such as caterpillars and grasshoppers, that can defoliate stockplants and also damage the growing shoot tips;

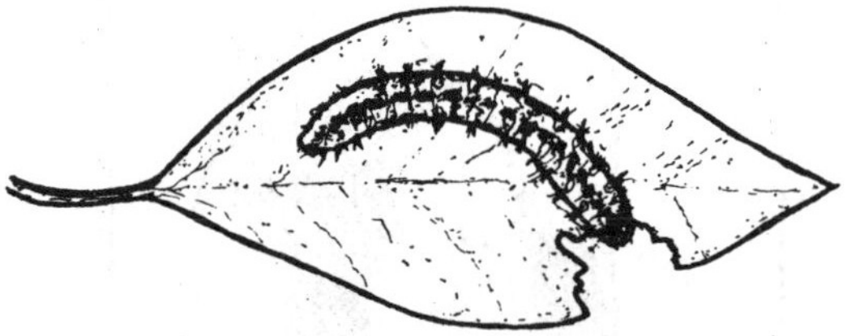

(3) **Red spider mites,** which can damage stockplants and cuttings;
(4) **Nest-building ants,** that could cause problems with stockplants;
(5) **Wood-eating termites,** which can damage propagators and wooden frames;
(6) **Root-feeding weevils,** that live in the soil.

What should I do?

(a) clear away material where insect pests may nest and breed.
(b) check the underside of leaves regularly for insect pests.
(c) where possible, remove the insects when you first see them, or spray with 5 drops of household detergent in 250 ml water.

(d) avoid having other host plants nearby. For example, the young stages of *Zonocerus variegatus* grasshoppers eat 'Siam weed' *(Chromolaena [Eupatorium] odorata)*, but later defoliate *Triplochiton scleroxylon*.
Some aphids need an *alternative host plant* to complete their life-cycle.

(e) use chemical sprays only when necessary, and not as a routine measure. Choose one that is recommended for the particular pest (A 63), use the dilutions advised, and ***follow the safety precautions carefully.***

Why shouldn't one use pesticides routinely?

(1) because they can leave toxic substances behind, which can harm humans and domestic animals. For this reason, some sprays such as *dieldrin* are no longer recommended for use;

(2) because the pests are more likely to develop *tolerance* to the chemical, which then becomes less effective.

(3) because sprays often kill the natural insect *predators* that keep down the pests.
(NOTE: biological control is preferable, if methods have been well tested.)

What chemicals are recommended?

There is a huge range of pesticides manufactured. Amongst those that are effective against several different kinds of insects are:
pirimicarb – especially against various kinds of aphids;
dimethoate – against aphids and red spider mites;
nicotine – against thrips, capsids, woolly aphids and some caterpillars.

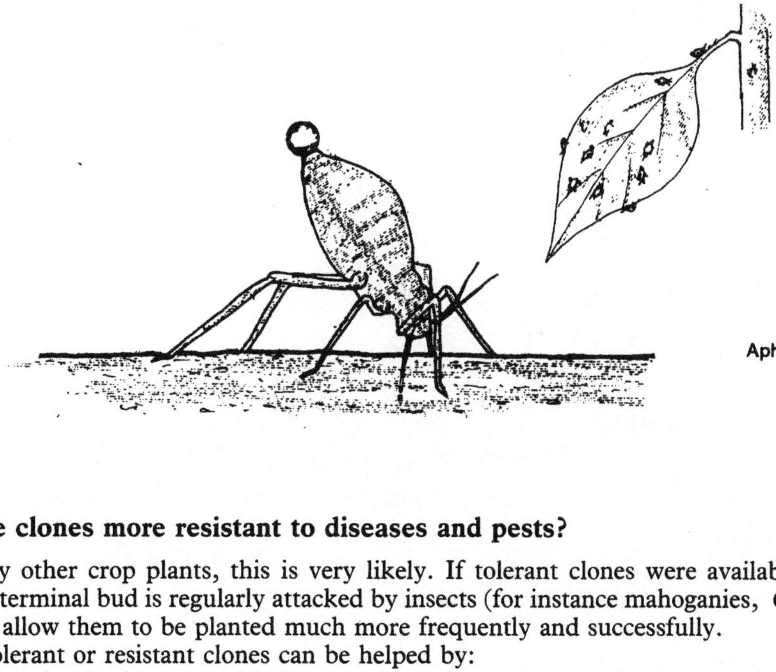

Aphids

Are some clones more resistant to diseases and pests?

Judging by other crop plants, this is very likely. If tolerant clones were available in species where the terminal bud is regularly attacked by insects (for instance mahoganies, *Chlorophora)*, this could allow them to be planted much more frequently and successfully.
Finding tolerant or resistant clones can be helped by:

(a) Choosing healthy trees (A 12), especially when others nearby have been attacked;

(b) Noticing whether stockplants and rooted cuttings of certain clones are frequently or seldom attacked;

At a later stage:

(c) Testing clones by actually inoculating them with a known disease, or putting the insect pest on them under controlled conditions;

(d) Breeding to combine tolerance with other desirable characters (Manual 2).

How about other pests?

Stockplants and cuttings might be damaged by:

(1) **nematodes** (eelworms) – tiny worm-like animals attacking roots.
(2) **rodents** – close any small holes that might allow mice to get into poly-propagators.
(3) **birds** – shade netting might be used (A 24, 33).
(4) **antelopes,** etc – fencing may be needed, and sometimes *pollarding* of stockplants (A 21).
(5) **domestic animals** – goats and hens are particularly destructive.

What other protection might be needed?

Don't forget that humans could damage your propagation work, for example by felling trees or collecting poles, burning areas for farming, harvesting leaves for eating and animal fodder, or through vandalism, including removing labels.

Care of cuttings A 53

– *potting up*

What is important about potting up rooted cuttings?

(1) It is a critical stage, at which it is easy to lose a lot of plants;
(2) It is not difficult, but a sound technique needs to be learnt.

Why is it a critical stage?

Because :

(a) the cutting is no longer in air of very high humidity;
(b) roots may easily get damaged;
(c) the new root system is usually small compared to the shoot;
(d) it has to grow into the new soil before it can function well;
(e) new leaves may be flushing, increasing the amount of water lost;
(f) the leading shoot may be growing, and can easily be broken off.

What sort of soil is suitable for potting?

Occasionally a local source can be used directly. However, it is generally better to mix various materials together, to make a potting soil which:

(a) has good aeration, to encourage root growth;
(b) has good drainage, but also retains moisture well;
(c) contains some organic matter and mineral nutrients;
(d) is easy to handle, and does not become hard or sticky;
(e) does not contain material infected with pests, diseases or weeds;
(f) does contain suitable *mycorrhizal fungi* (Manual 3), and (in the case of some *leguminous* tree species) *nodulating bacteria.*

Why do I need a special potting soil?

(1) Because using the nursery soil usually gives poor results;
(2) Because a tree in a pot makes greater demands on a limited amount of soil for water, air and nutrients than one rooted in the ground;
(3) Because some species will otherwise not thrive (for example *Cedrela* if drainage is poor);
(4) To give better survival and more uniform growth in the nursery (Manual 3);
(5) To form a good root system which will help in field establishment (Manual 5).

What can I mix together?

(1) **Forest top soil** – collect in relatively undisturbed forest, especially the surface layer that is a darker colour – good for (c) and (f) above;
(2) **Black soil** from decayed rubbish, but watch out for (e) and broken glass;
(3) **Coarse sand** (0.2-2.0 mm) or **grit** – especially if you need to improve (a), (b) and (d);
(4) **Sawdust or shredded bark** (well 'weathered') – for (b) and (c);
(5) **Crop residues,** such as groundnut shells, rice husks or coconut fibre – for (c);
(6) **Inoculum** for favourable mycorrhizas or nodules, if available – useful (and occasionally essential) for (f) (Manual 3);
(7) **Fertilisers or trace elements:**
- where there is a particular shortage of nutrients in local materials;
- when rapid growth is needed;
- for trees grown for research as pot plants, or in the field;

(8) **Sub-soil** may occasionally be used to bulk up the quality of potting soil, but ***not*** if it is clayey, hard or full of stones.

How much of each should I use?

This depends very much on the type of materials available. However, you could start with:- one-third of 1 or 2 above; one-third of 3; one-third of 4 or 5. It is a good plan to try out several contrasting potting mixes, using uniform batches of rooted cuttings or seedlings.
Sieve out any large pieces, and mix the potting soil well, adding water if needed to make it moist enough that it just holds together when squeezed in your hand.
(NOTE: Don't make it too wet!)

When should I pot up cuttings?

When they have at least 2 cm total length of roots, but before any of them get longer than about 4 cm. This stage can be reached in as little as 10 days, but it may take 3 – 6 weeks, and sometimes more.

How can I avoid damaging the roots?

(1) By potting up in a moist, shady place, protected from wind;
(2) By getting the pots, potting soil, labels, pencil etc, ready first;
(3) By digging the cuttings up gently with a small, flat piece of wood;
(4) By digging up a few at a time, so they don't dry up;
(5) By holding the cutting by the stem, and leaving any rooting medium still attached ***(Not too deep or too shallow);***

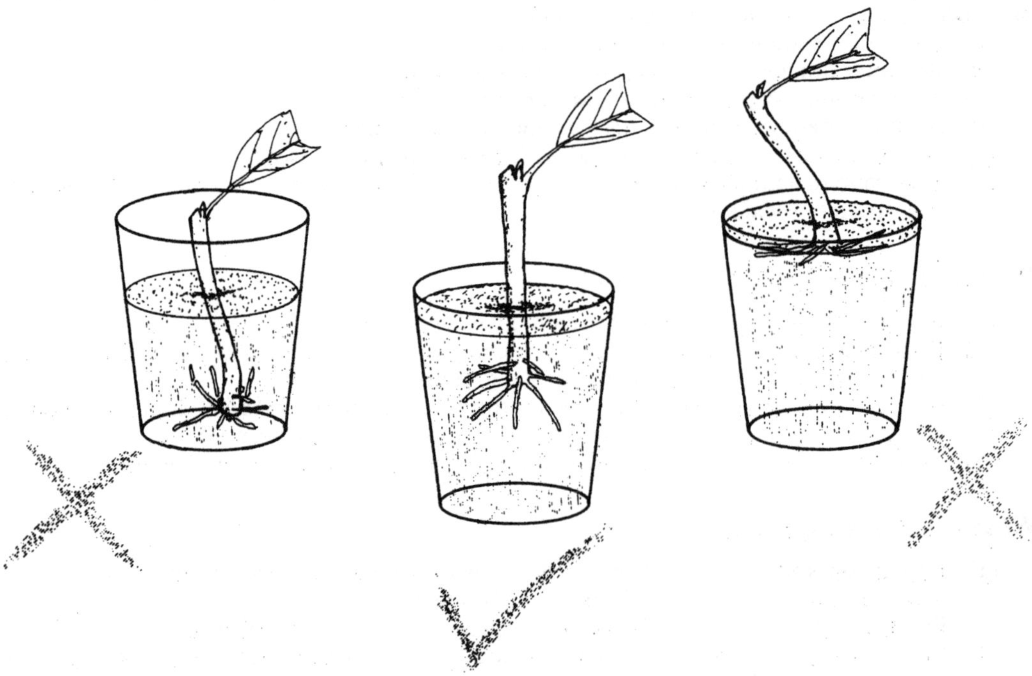

(6) By putting some of the potting soil into the pot first, before inserting the cutting. Then add some more soil, and firm it down with the fingers without breaking the roots or pulling them off the stem; ***(This is the most critical stage!)***
(7) By rooting in *bio-degradable* containers, which rot away after a few months.
(Note: don't let them dry out or they may restrict root growth.)
The rooted cuttings, still in the container, can then be potted on into the normal size of pot without disturbing the root system (A 36).

How can I speed up the work?

(a) by training a team to work well together;
(b) by making a simple funnel to get the potting soil quickly into polythene pots;
(c) by using a bonus scheme based on successful potted cuttings.

Where should I put the pots afterwards?

They need to go straight into shady, humid conditions again. A good place is an empty propagator, or you could make a movable cover lined with polythene sheeting.

But don't they have to get used to ordinary conditions?

Yes, but this needs to be done gradually. If you go too fast, wilting is likely (A 34), which will slow growth, and may damage or kill the young plant. For further details of *weaning and hardening* see A 54.

Do I water them in?

No, it is much better to use potting soil that is already moist enough, and to rely on high humidity to check evaporation. When the soil is beginning to dry, moderate and careful watering is needed (A 51).

What else is needed?

Write some labels, showing the batch number, and date of potting, and fill in the record sheet (A 67). It is better to go to the trouble of labelling each plant than to get the clones mixed up.

This page has been left blank for field notes.

Care of cuttings A 54
– *weaning and hardening*

What is weaning?

Gradually accustoming potted cuttings to grow under ordinary nursery conditions.

How is it done?

By decreasing, step by step, the additional humidity (or mist) that is needed for rooting the cuttings.

What's the best way of doing that?

For poly-propagators, you could work out stages such as these:
Stage 1 – newly potted cuttings remain in poly-propagator or under polythene cover;
Stage 2 – covers opened up during the night;
Stage 3 – covers opened up night and day, except in very hot spells;
Stage 4 – weaned plants moved to nursery beds, still under shade;
Stage 5 – reduce the shading (light-demanding species).

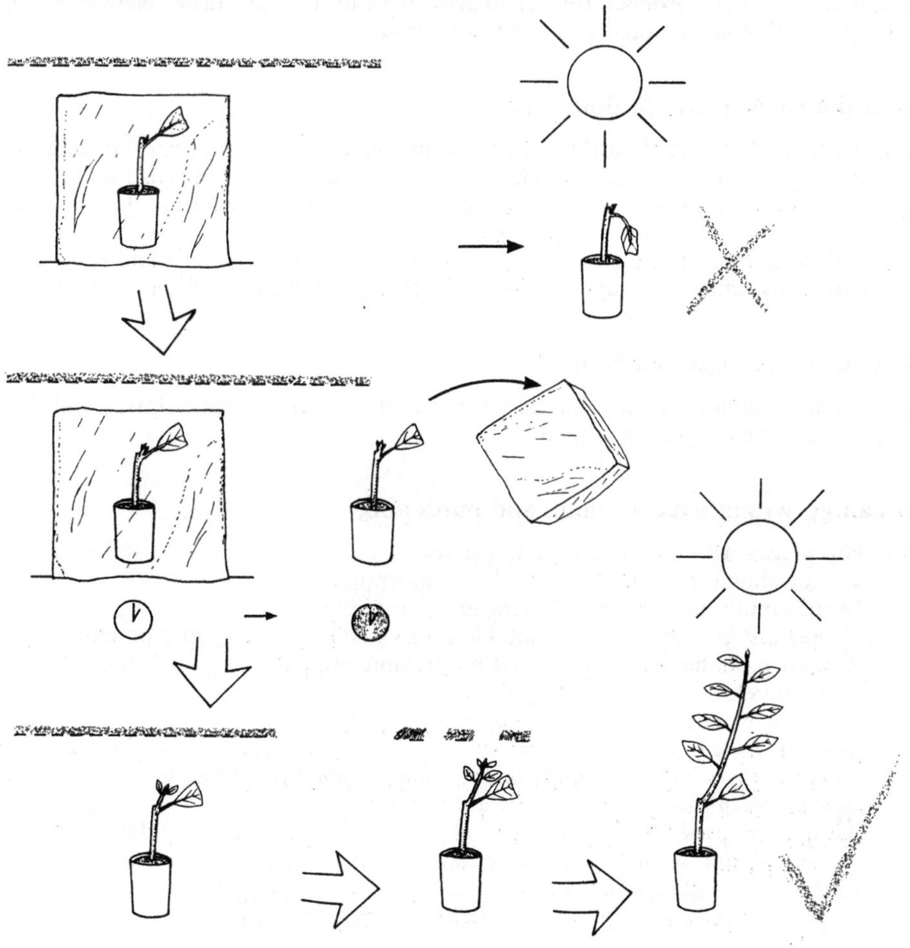

How long should each stage be?

Try weaning with one week each for Stages 1 – 3. However, not all tree species react the same, and longer may be needed for stage 1 in hot, dry weather.
If weaning is done too fast, plants may wilt, causing a check to growth and perhaps damage or deaths (see A 34, 51).
If it is done too slowly, the trees may make a lot of very soft new growth, which they cannot maintain with the roots they have.

What about weaning from mist?

The same stages could be given under polythene covers, or alternatively the misting programme could be altered each week to give shorter, less frequent bursts, and no mist at night.

How about weaning from micropropagation?

The plantlets are very small, and have been growing in low light, so they need a longer period to acclimatise to nursery conditions. Follow the suggestions in sheet A5, and don't rush the early stages!

What is hardening?

This is the gradual reduction of shading to accustom the plants to stronger sunlight. It is similar to the hardening of seedling stock before planting, but can take rather longer because the original conditions were very humid.
With *shade-bearing* species, retain some shade in the nursery, as for seedling plants. *Light-demanding* trees can be hardened further to grow with little or no shade. New leaves may have to be produced that are adapted to a drier environment.

What is the easiest way of doing this?

If there are only a few rooted cuttings, they can be moved, in several stages, from a very shady place to full sun. If there are large batches, it can be easier to alter the shading progressively:

(a) **with palm-leaf shade:** as the fronds dry, more light will penetrate. Finally thin out the shade, and then remove it entirely;

(b) **with shade-cloth:** use two grades, such as those which cast 60% and 30% shade. Start with both, which gives about 72% shade, then go in turn to 60%, 30% and full sunlight.

How long should hardening take?

If possible, allow about a month for each step, so that more hardened leaves and stem tissue can have time to develop at each stage.

What can go wrong with weaning and hardening?

(1) **The plants wilt: this could be because:**
(a) they should have been watered that morning;
(b) the conditions have been changed too rapidly;
(c) they are growing in an unsuitable potting soil, or were badly potted;
(d) their roots have been damaged by previous drought or waterlogging;
(e) too much fertiliser has been applied;
(f) they have been attacked by a disease or pest (A 52);
(g) weeds have been allowed to grow in the pots, and these are absorbing water rapidly.
(NOTE: slight mid-day wilting in hot, sunny weather is normal, while in some plants the angle at which leaves or leaflets are held can alter.)

(2) **Plants stop growing:** besides 1(b-f) above, possible reasons include:
(a) their natural growth habit is by repeated flushing and dormancy;
(b) they are shade-bearing species reacting to bright sunlight;
(c) they are short of important nutrients or trace elements (A 27).

What else is important?

Stopping the plants from rooting through into the ground below. This is a common reason for poor field establishment, which originates in the nursery. It happens because:

(1) plants often grow *better* when their roots can explore a greater volume of soil;
(2) this makes them more tolerant of poor potting soils and to lax nursery procedures;
(3) however, such plants may make rather few fine roots inside the pot;
(4) when they are moved, their main roots are broken, and they often wilt irreversibly and die.

How can I stop them rooting through?

(a) by using polythene pots with adequate drainage holes in the sides, ***but not in the bottom;***
(b) by standing the pots on polythene sheeting *(but avoid puddles);*
(c) by putting them on a concrete surface;
(d) by standing them on broken stones, so that emerging roots dry up;
(e) by making benching to keep them clear of the ground;
(f) by moving them frequently, and breaking or cutting emerging roots.

What else do I need to know about the nursery stage?

(1) Information and advice on sources of seed and the growing of seedlings. This is to be the subject of **Manual 2 – *Raising Seedlings of Tropical Trees;***
(2) Similarly on the planning and running of tree nurseries, and plant nutrition. This will be covered in **Manual 3 – *Growing Good Tropical Trees for Planting.***

Can I now be sure my cuttings will grow and establish well?

They will definitely have a very good chance of doing so. However, there are still two stages at which all the work could fail in its objective:

(a) planning and preparation of the site have to be thought through. This will be examined in **Manual 4 – *Preparing to Plant Tropical Trees;***
(b) planting and weeding have to be well done, so that your trees can establish successfully. This will be explored in **Manual 5 – *Planting and Establishment of Tropical Trees.***

This page has been left blank for field notes.

Check-lists, sources and records A 60
– *some key requirements for tree-planting projects*

(A) Organisational and administrative

(1) Sufficient **forward planning and continuity** of the project to provide staff with reasonable confidence about the future.
(2) More than one graduate on the **full-time staff** of the project, and local **counterparts** working full-time with each expatriate member of staff.
(3) Adequate **funding** through a straightforward system, allowing easy budgeting, financial control, and the purchasing of important items locally and abroad.
(4) Enough **administrative staff**, where appropriate bilingual, to deal efficiently with the office work and routine running of the project, so that the technical and scientific staff are relieved of part of this work-load.
(5) Successful **integration** of research, plant propagation, tree establishment, social concerns and ongoing management aspects of the project.
(6) Good two-way **communication** and other links with the local community.
(7) Effective safeguarding of the **land, buildings, installations and trees** against encroachment, fire, theft and vandalism.
(8) Well-maintained **vehicles**, controlled by the project, and protected at night.
(9) Reliable control, maintenance and care of **equipment, tools and materials**.

(B) Training and collaboration

(1) Emphasising the **continuing training** of staff and workers, including appropriate travel to other parts of the country and abroad.
(2) Timing of travel for **longer periods of training** so as to minimise disruption of the project.
(3) Running of **short, practical courses** that are designed to spread the approaches, know-how and enthusiasm generated by the project; and collaborating with people locally in identifying **present and future needs**.
(4) **Publishing** the project findings in scientific, technical and general journals; and in the form of practical advice to potential users.
(5) Arranging **discussions and relevant collaboration** with Universities, Colleges, Institutes and other Projects; in such fields as forestry, agriculture, agroforestry, biology, tropical land use and rural development; including the training and interchange of research students.

(C) Growing and planting trees

(1) Wherever necessary, improving existing nursery techniques, so that plenty of **good nursery plants** can be produced, that can establish well when planted out, and grow into trees of above average **form, usefulness and yield**.
(2) Ensuring effective and reliable systems for **watering and care** of nursery pot plants.
(3) Recognising how important it is to have **good root systems** *inside* the pot, and not in the ground beneath the pot.
(4) For clonal tree-planting, working out appropriate **management systems for the stockplants** and **watering regimes for the poly-propagators** (or the best misting regimes for automatic mist-jets), in both rainy and dry seasons.
(5) Providing year-round care of the trees being propagated, and foreseeing breakdowns before they occur.
(6) Doing **experiments** to try out the techniques on local species, and to clarify the factors behind nursery and establishment problems.
(7) Gaining access to and control over **suitable ground** that is designated and kept for trials, plantations, demonstrations and other types of tree planting.
(8) Using long-lasting **labelling** methods, and maintaining good **records**.
(9) Ensuring regular **weeding and climber cutting** until trees are established.

This page has been left blank for field notes.

Check-lists, sources and records A 61
– *check-list of problems in growing good trees*

Stage or feature	Problem	Possible solutions	Alternative strategies
(A) SEEDS			
1 – PROVENANCE *unacceptable* → *suitable* ↓		collect or import seed from other sources	
2 – PARENT TREES *unacceptable* → *suitable* ↓		collect from several parent trees of good form, grouped near each other, not closely related	
3 – FLOWERING *intermittent* → *regular* ↓		use stored seed or wildings; apply flower induction treatments (if available – see Manual 2)	use cuttings
4 – FRUITING *irregular* → *regular* ↓		use stored seed or wildings	use cuttings
5 – QUALITY OF SEED *poor* → *good* ↓		collect early, treat against pests and diseases	use cuttings
6 – QUANTITY OF SEED *insufficient* → *enough* ↓		collect by climbing	multiply by cuttings
7 – VIABILITY IN STORAGE *poor* → *prolonged* ↓		sow at once; or improve drying &/or storage methods	
8 – SEED DORMANCY *a problem* → *prompt germination* ↓		apply appropriate technique to break the type of dormancy present (see Manual 2)	
9 – YOUNG SEEDLINGS *die young* → *thrive* ↓ *[go to (C)]*		check soil, shade, watering, spray for damping-off fungi, protect from small animals	

Stage or feature	Problem	Possible solutions	Alternative strategies
(B) LEAFY CUTTINGS			
1 – WILTING	*a problem* → *rare* ↓	take cuttings early, check high humidity is maintained in bags & propagator, check shading	
2 – LEAF-DROP	*common* → *unusual* ↓	check as (B) 1; use younger leaves, handle cuttings more carefully	use pre-treatment
3 – ROTTING	*a problem* → *rare* ↓	check rooting medium, watering, shading; use fungicide	try leafless cuttings *in situ*
4 – SLOW ROOTING	*typical* → *uncommon* ↓	check stockplant management, rooting environment, use auxins	
5 – SINGLE ROOTS	*usual* → *unusual* ↓	use auxins, try longer cuttings	
6 – INHIBITED ROOT GROWTH	*typical* → *atypical* ↓	try less shade, weaker auxin; check fertilising of stockplants and cutting condition	
7 – MULTIPLE SHOOTS	*common* → *uncommon* ↓	remove tips, leaving best; use less branchy clones	try slit half-cuttings with single bud
8 – WEANING DEATHS	*many* → *few* ↓	dig up and handle cuttings more carefully, wean more slowly; check soil, watering and fertilisers	
(C) YOUNG POTTED PLANTS			
1 – SURVIVAL	*low* → *high* ↓	check soil, watering, shading, pests & diseases, fertilisers; change container	check need for mycorrhizas
2 – INHIBITED GROWTH	*pronounced* → *slight* ↓	check as (C) 1, and try to avoid transplanting and weaning stress – check root systems	
3 – GROWTH RATE	*shoots too tall* → *height acceptable* ↓	alter timings, reduce fertilisers	try cuttings at a different time
(D) SUCCESSFUL PRODUCTION OF SUFFICIENT, GOOD PLANTING STOCK			

Stage or feature	Problem	Possible solutions	Alternative strategies
(E) PLANTING & ESTABLISHMENT			
1 – SURVIVAL *low* → *good* ↓		check condition of planting stock (especially root systems), transport & handling, timings re rains, soil, planting techniques, weeds, pests & diseases	prevent rooting through in nursery; try mycorrhizas
2 – INHIBITED GROWTH *general* → *uncommon* ↓		check as (D) 1, and (where relevant) consider mulch, fertiliser, legume shade tree, critical stage watering	change container
(F) SUCCESSFUL TREE PLANTING			
(G) LEAFLESS CUTTINGS (planted in situ, *directly in place)*			
1 – SURVIVAL & GROWTH *poor* → *good* ↓		check length, diameter and age of cuttings; check handling & timing	try leafy cuttings
2 – MULTIPLE STEMS *common* → *rare* ↓		remove lower buds and branches; use less branchy clone	
(H) SUCCESSFUL TREE PLANTING			

(See sheet A 2 for further explanation of this check-list, and the Contents list for sheets with more information on the suggested actions.)

This page has been left blank for field notes.

Check-lists, sources and records A 62
– *information on vegetative propagation*

Vegetative propagation & tree improvement:

Baker, F.W.G. (Ed.) (1992). *Rapid propagation of fast-growing woody species*. C.A.B. International, Wallingford, Oxon, Britain.

Davis, T.D., Haissig, B.E. & Sankhla, N. (Eds.) (1988). *Adventitious root formation in cuttings*. Advances in Plant Sciences, Vol 2, Dioscorides Press, Portland, Oregon, USA.

Jackson, M.B. & Stead, A.D. (Eds.) (1983). *Growth Regulators in Root Development*, Monograph 10, Br. Plant Growth Regulator Group, Wantage, Britain.

Ladipo, D.O., Leakey, R.R.B. & Grace, J. (1991). Clonal variation in a four year old plantation of *Triplochiton scleroxylon* K. Schum. and its relation to the Predictive Test for Branching Habit. **Silvae Genetica, 40,** 135-140.

Leakey, R.R.B. (1983). Stockplant factors affecting root initiation in cuttings of *Triplochiton scleroxylon* K. Schum., an indigenous hardwood of West Africa. **J. hort. Sci., 58,** 277-290.

Leakey, R.R.B. (1985). The capacity for vegetative propagation in trees. Pp. 110-133 in: *Attributes of Trees as Crop Plants*, edited by M.G.R. Cannell & J.E. Jackson, Institute of Terrestrial Ecology, Abbots Ripton, Huntingdon, Britain.

Leakey, R.R.B. (1987). Clonal forestry in the tropics – a review of developments, strategies and opportunities. **Commonwealth Forestry Rev., 66,** 61-75.

Leakey, R.R.B. (1990). The domestication of tropical forest trees by cloning: a strategy for increased production and for conservation. Pp. 22-31 in *Fast Growing Trees and Nitrogen Fixing Trees*, edited by D. Werner & P. Müller, Gustav Fischer Verlag, Germany.

Leakey, R.R.B. & Ladipo, D.O. (1987). Selection for improved tropical hardwoods. Pp. 229-242 in *Improving Vegetatively Propagated Crops*, edited by A.J. Abbott & R.K. Atkin, Academic Press, London, Britain.

Leakey, R.R.B & Longman, K.A. (1986). Physiological, environmental and genetic variation in apical dominance as determined by decapitation in *Triplochiton scleroxylon*. **Tree Physiology, 1,** 193-207.

Leakey, R.R.B., Chapman, V.R. & Longman, K.A. (1982). Physiological studies for tropical tree improvement and conservation. Some factors affecting root inititation in cuttings of *Triplochiton scleroxylon* K. Schum. **Forest Ecol. & Management, 4,** 53-66.

Leakey, R.R.B. and others (1990). Low-technology techniques for the vegetative propagation of tropical trees. **Commonwealth Forestry Rev. 69,** 247-257.

Libby, W.J. & Rauter, R.M. (1984). Advantages of clonal forestry. **Forestry Chronicle, 60,** 145-149.

Pochet, P. (1987). Le bouturage du caféier Robusta/*Robusta propagation by cuttings*. Administration Générale de la coopération au Développement (AGCD), 5, place du Champ de Mars, 1050 Bruxelles, Belgium.

Tissue culture & micropropagation:

Biondi, S. & Thorpe, T.A. (1982). Clonal propagation of forest tree species. Pp. 197-204 in: *Tissue Culture of Economically Important Plants*, edited by A.N. Rao, Comm. on Sci. & Tech. in Devel. Countries/Asian Network for Biol. Sci., Singapore.

Bonga, J.M. & Durzan, D.J. (Eds.) (1982). *Tissue culture in forestry*, Martinus Nijhoff/ W. Junk, The Hague, Netherlands.

Jones, O.P. (1983). *In vitro* propagation of tree crops. Pp. 139-159 in: *Plant Biotechnology*, edited by S.H. Mantell & H. Smith, Soc. exptl Biol., Seminar Series 18, Cambridge Univ. Press, Britain.

Sommer, H.E. & Caldas, L.S. (1981). *In vitro* methods applied to forest trees. Pp. 349-358 in *Plant Tissue Culture: methods and applications in agriculture*, edited by T.A. Thorpe, Academic Press, New York, USA.

Staritsky, G. & van Hasselt, G.A.M. (1980). The synchronised mass propagation of *Coffea canephora* in vitro. pp. 597-602 in: *Proc. 9th internat. Colloquium on Coffee*, Vol. 2, London, Britain.

Nursery manuals:

Dirr, M.A. & Heuser, C.W., Jr. (1987). *The reference manual of woody plant propagation*. Varsity Press, Inc., P.O. Box 6301, Athens, Georgia, USA.

Duryea, M.L. & Landis, T.D. (Eds.) (1984). *Forest Nursery Manual: production of bareroot seedlings*. Martinus Nijhoff/W. Junk, The Hague, Netherlands.

Liegel, L.H. & Venator, C.R. (1987). *A technical guide for forest nursery management in the Caribbean and Latin America*. Forest Service, U.S. Dept. of Agric., Gen. Tech. Rep. SO-67.

Napier, I. (1985). *Técnicas de viveros forestales con referencia especial a Centroamérica*. Publicación Miscelanea No. 5, Escuela Nacional Ciencias Forestales, Siguatepeque, Honduras, Central America.

Napier, I. & Robbins, M. (1989). *Forest seed and nursery practice in Nepal*. Nepal/UK Forestry Research Project, Dept of Forestry & Plant Research, Babar Mahal, P.O. Box 3339, Kathmandu, Nepal.

Willan, R.L. (1985). *A guide to forest seed handling, with special reference to the tropics*. FAO, Rome, Italy & DANIDA Forest Seed Centre, Humlæbek, Denmark.

General information on growing trees:

Aumeeruddy, Y. & Pinglo, F. (1989). *Phytopractices in tropical regions*. Man and the Biosphere Program, UNESCO, Paris/Institut de Botanique, Montpellier, France.

Carter, E.J. (1987). *From seed to trial establishment*. DFR User Series No. 2, Division of Forest Research, Commonwealth Scientific & Industrial Research Organisation (CSIRO), P.O. Box 4008, Yarralumla, ACT 2600, Australia.

Dupriez, H. & de Leener, P. (1983). *Jardins et vergers d'Afrique*. English translation (1989): *African Gardens and Orchards*. Macmillan/Terres et Vie, 13, rue Laurent Delvaux, 1400 Nivelles, Belgium.

Hartmann, H.T. & Kester, D.E. (1983). *Plant Propagation: principles and practices*, 4th edn, Prentice Hall, New Jersey, USA.

Longman, K.A. & Jeník, J. (1987). *Tropical Forest and its Environment*, 2nd Edn, Longman Scientific & Technical, Harlow, Britain & John Wiley, New York, USA.

Tinus, R.W. & McDonald, S.E. (1979). *How to grow tree seedlings in containers in greenhouses*. Forest Service, U.S. Dept. of Agric., Gen. Tech. Rep. RM-60.

Wilson, J. & Leakey, R.R.B. (1990). Repairing the damage: re-establishing trees in the tropics. **The Crown Agents Review**, London, **1**, 14-20.

Check-lists, sources and records A 63
– *Some sources of chemicals and materials*

(A) Examples of suppliers of chemicals, ***some with agents in the tropics:***

(1) **Auxins for rooting cuttings:**

Dip 'N Grow – contains 1.0% Indole-3-butyric acid (IBA) plus 0.5% 1-napthaleneacetic acid (NAA) in alcohol: from *Alpkem Corporation, Clackamas, Oregon 97015, USA.*

Hormo-Root – contains 0.1-4.0% IBA plus Thiram (a fungicide) in talc (inert powder); from *Hortus Products Co., P.O. Box 275, Newfoundland, New Jersey 07435, USA.*

Rootone – contains 0.057% IBA plus a total of 0.113% of 3 chemical derivatives of NAA, plus Thiram (fungicide), in talc; from *Union Carbide Agricultural Products Co., Inc, P.O. Box 12014, T.W. Alexander Drive, Research Triangle Park, North Carolina 27709, USA.*

Seradix – contains 0.1-0.8% IBA in talc; from *Rhone-Poulenc, Agricultural Centre, 14-20 Rue Pierre Baizet, 69009 Leon, France* [Tel: (33) 7229 255]

Strike – contains 0.2-0.3% NAA plus Captan (fungicide): from *Pan Britannica Industries (pbi), Britannica House, Waltham Cross, Herts, EN8 7DY, Britain.* [Tel: 0992 23691; Telex: 23957; Fax: 0992 26452].

(2) **Pure chemicals for micropropagation and research are available from many firms, including:**

Cyanamid, *C House, Fareham Road, Gosport, Hants, PO13 0AS, Britain* [Tel: 0329 224000; Telex 86173; Fax: 0329 220213].

Mackay & Lynn Ltd, *2 West Bryson Road, Edinburgh, EH11 1EH, Britain* [Tel: 031-337 9006] *(Also laboratory equipment and sundries.)*

Sigma Chemical Co Ltd, *Fancy Road, Poole, Dorset BH17 7TG, Britain* [Tel: 0202 73314; Telex: 418242; Fax: 0202 715460 *(Group 2/3)]*.

(3) **Pesticides, fungicides, weedkillers, fertilisers, etc are also widely available. Main suppliers include:**

Du Pont (UK) Ltd, *Agricultural Products Department, Wedgwood Way, Stevenage, Herts, SG1 4QN, Britain.* [Tel: 0438 734000; Telex 825591].

Fisons Horticulture Division, *Paper Mill Lane, Bramford, Ipswich, Suffolk, IP8 4BZ, Britain.* [Tel: 0473 830492; Telex: 98240].

Imperial Chemical Industries (ICI), *Woolmead Walk, Farnham, Surrey, GU9 7UB, Britain.* [Tel: 0252 733919 (horticultural products); 0428 644061 (agrochemicals).

Pan Britannica Industries (pbi), *Britannica House, Waltham Cross, Herts, EN8 7DY, Britain.* [Tel: 0992 23691; Telex: 23957; Fax: 0992 26452].

Note that ***Benomyl*** = **Benlate fungicide**, and is a wettable powder that can be used for sprays, dips or soil drenches. It acts *systemically*, by being taken up by the plant and spreading inside it.

(B) Horticultural materials & equipment: for example from:

Grangewood Plastic Packaging Ltd, *Essex House, Jutsoms Lane, Romford, Essex, RM7 0ER, Britain* [Tel: 0708 725911; Telex: 8951426; Fax 0708 728 677] *(Polythene sheet and bags).*

Kerrypak Ltd, *Longbrook House, Ashton Vale Road, Bristol, BS3 2HA, Britain* [Tel: 0272 669 684, 0272 662 455; Telex: 444 277; Fax: 0272 231 251] *(Plastic shadecloth).*

LBS Polythene, *Cotton Tree, Near Colne, Lancs BB8 7BW, Britain.* [Tel: 0282 871777; Fax: 0282 869850] *(Wide range of products).*

Right Rain, *Stag Business Park, 164-166 Christchurch Road, Ringwood, Hampshire, BH24 3AS;* [Tel: 0425 472 251; Telex: 41206; Fax: 0425 472 258] *(Mist propagation equipment).*

Spencer-Lemaire Industries, Ltd, *10310 112 Street, Edmonton, Alberta, Canada, T5K 1N1* [Tel: (403) 426 3203] *(Rootrainers).*

(Note: when calling from outside Britain, omit the first zero from telephone and fax numbers.)

Check-lists, sources and records A 64
– *record-keeping and labelling*

Why keeping records is important:

Tropical trees are not plants with short life-cycles, like many food crops. Making records is therefore **especially important**, for instance, to:

(a) write down basic information that will soon be forgotten or confused;
(b) inform someone else about the trees;
(c) keep track of which clone a tree belongs to, and avoiding muddle if labels are lost or become indistinct;
(d) be able to find the original coppice stump or seedling from which a clone was started;
(e) keep stock of the potted trees available for planting;
(f) know which clones and batches of trees are doing best;
(g) see how to improve the quality and rate of production of nursery stock.

Some hints about making records:

(1) make them on the ***same day*** as the work is done. Draw a simple sketch map to show the positions of coppice stumps, planted stockplants, cuttings being rooted, experimental lay-outs, etc.
(2) try to make them *user-friendly*, so that you or anyone else can understand them after some time has passed;
(3) when similar information will often be recorded, photocopy one of the sheets in the Manual (A 66-68), change it to suit, or design your own;
(4) Always put down the date and the clone or batch numbers;
(5) Keep extra copies of essential information in case of loss of files, note-book or labels.

Different types of label:

For short-term purposes (less than a year), many types of label are suitable, for example those made of plastic, wood or metal.
However, it is difficult to find labelling that will last a long time in the tropics, because of the effects of bright sun, heavy rain, termites, weed growth, etc. Plastic can become brittle in the sun, wood may rot and metal tarnish.

Writing the information:

(a) reduce the amount to be written to a minimum, ***but remember to keep an explanation in the records;***
(b) the **general** information, like date, aim or experiment, species, etc, can go on a few (larger) labels, for example one at the beginning and one at the end;
(c) the **specific** information, like clone, type of cutting, treatment, etc, needs to be written for each tree (or sometimes batch of trees).

What to write with:

(1) **Pencil** – with a good HB, B or 2B pencil, pressing hard on the label, writing on metal, plastic or painted wood can remain legible for several years;
(2) **Waterproof felt-tip pens** – writing done with these can last for up to 1-3 years on wooden labels, and also if written on smooth-barked trees when the surface is dry and clean. These indelible markers are quick and easy, and can also be used on leaves (A 42). On plastic and metal labels, the writing tends to fade quickly, especially during propagation;

(3) **Paint** – slower and more cumbersome, but lasts well if an *exterior quality* paint has soaked into the bark or wooden label;
(4) **Numbers on concrete posts** – this is the longest lasting method, but it takes more time. It needs a set of *reversed* numbers and letters that can be re-used after they have left their imprint in the concrete.

Where to attach the labels:

(a) **pushed into the pot** – best when the plant is too small to attach a label; but note that they can fall out, or get mixed up during repotting or planting out;
(b) **attached to small plants** – pass a plastic *twist*, a small piece of thin wire or strong twine through a hole in the label and fasten round the main stem below a branch. ***Note: make a loop that is large enough for the plant to grow in diameter, but not big enough for it to come off.***
(c) **attached to larger trees** – as soon as possible, transfer the loop and label to a strong branch, or make a much larger loop of thicker wire (to lie on the ground around the main stem) for planted saplings;
(d) **set in the ground** – labels can be nailed to stakes made of *durable* wood, and driven into the ground; concrete marker posts can be set in a hole with stones, and a sand/cement mixture poured on.

Problems with labels:

(1) branches holding labels may die and be shed;
(2) as the main stem grows, labels can become detached, or strangle the tree;
(3) writing on the trunk separates and gets fainter as the tree grows in girth;
(4) labels in wet places can get covered with dirt or small green plants like *algae* or mosses. Clean them gently, not to rub off the writing;
(5) certain animals (for example porcupines and birds) may damage labels;
(6) vandals sometimes remove labels, so it can be best not to display them too prominently;
(7) on the other hand, a rapid change of appearance of a site may make it hard to find the trees again, so draw a sketch-map, and standardise the position of the label.

Using the records

The best-kept records are of little use if they are not used! Here are two suggestions:

(A) Arrange your records in a file or notebook that you can easily carry round with you.

(B) Make a habit of checking up at the same time each week on stockplants, propagation conditions, numbers of plants in each clone or batch, etc, *using your records to do this*, and updating them as you go.

(A) Need for scoring methods

Scoring is a useful way of getting a rapid, general view of a situation in biology. Use it before embarking on long and detailed measurements that may or may not be appropriate, and especially when the features to be assessed are difficult or impossible to record by measurement or counting; for example when:

(1) the feature is primarily **qualitative**, such as the presence or absence of visible roots on a cutting, or the incidence of forking in the main stem of a clone;
(2) the feature is rather **subjective**, such as the branching habit of a tree;
(3) the feature cannot be accurately measured **without killing** the plant organs in question, as for example with the dry weight of roots on a cutting;
(4) the trees are **too big** for easy measurement;
(5) there are **too many** items to count.

(B) Weaknesses of scoring methods

(1) it can be difficult to standardise the categories;
(2) intervals between categories are not necessarily equal;
(3) bias is harder to avoid;
(4) after a time, the brain may refuse to carry on scoring without a rest;
(5) some statistical tests might be impossible, or could be misleading.

(C) Features of useful scoring methods

The main aims when choosing an appropriate scoring method are to minimise the weaknesses and achieve a valid and useful assessment without undue difficulty or delay. Some hints are:

(1) Do the scoring with at least one other person – difficult features may need three or four observers. Discuss the features together before scoring independently;
(2) If possible, look through the material first in order to gauge the range of variation to be expected in the feature(s) to be scored; and to discover whether the trees have yet reached a suitable stage of development;
(3) Choose between 5 and 10 categories which cover the whole range of the feature, and give each category a number *(for example 5 = very straight stem; 1 = very crooked stem);*
(4) A short trial run is often a help, as you may want to modify the categories;
(5) Consistency is more important than the level chosen;
(6) Don't try and score too many different features at the same time;
(7) To reduce bias in experiments, it may be better to do the scoring without knowing the treatment, variety, clone, etc.

(D) Analysis of scored data

(1) **Chi-square tests** are very quick to calculate, and give a simple estimate of the significance of qualitative differences such as the presence or absence of a feature. Where features have been scored into several categories, but the frequencies in some are too low for valid analysis, they can be grouped (for example) into categories 0 + 1, and categories 2 – 5, and the two groups analysed by a chi-square test.
(2) **Analysis of variance** can be applied to scored data, provided that:
a) the variation is continuous and the intervals between categories are reasonably even;
b) an appropriate *transformation* is used when the data is not normally distributed, or is expressed as a percentage. If there are many zero values, it may be sensible to compare the presence or absence of the feature by a chi-square test, and confine the analysis of variance to the cases where the feature is present.

Transformations may also be relevant when the categories are non-linear (for example, with numbers of roots in categories of 0, 1, 2-5, 6-14, 15+).

If variation between independent observers is included in the analysis, the error mean square (variance estimate) may be reduced.

(E) Summary

Used with judgement, scoring methods can provide a rapid and useful complement to more precise and quantitative measurements. They are especially appropriate when the features do not lend themselves to easy measurements, or when time is short. Although the data obtained are only semi-quantitative, it may be possible to carry out valid statistical tests of significance.

Check-lists, sources and records – *record sheet for clone numbers* A 66

Species		**Provenance**		**Clone number**	

Whereabouts of original plant: Altitude Approximate age

Features of original plant: Date selected Approx. height

diameter

stem straight?	stem cylindrical?	forking?
live branch angle?	live branch size?	dead branch shedding?
live crown depth?	live crown evenness?	health and other features?

Cuttings taken:	type	number taken	number set
		on (date)	on (date)
		number rooted	number potted
		on (date)	on (date)

Planted as stock-plants:	place	position
	number	date

Planted as trees:	place	position
	number	date

Notes:

This page has been left blank for field notes.

SPECIES/*ESPECE:*

Date: / / Identity/*Identité no.*

ORIGIN OF STOCKPLANTS/*ORIGINE DES PLANTES RESERVEES COMME SOURCE:*

Country/*pays* – Provenance/Land race –

Exact locality/*Localité exacte* – Altitude –

STOCKPLANTS/*PLANTES RESERVEES COMME SOURCE:*

Location/*Localisation* – Clone no(s) –

Height cuttings taken/*Hauteur relative qu'on a pris les boutures* – m.

No. of stockplants used/*No. des plantes utilisées comme source* –

Previous treatment of stockplants/*Traitement antérieur des plantes* –

CUTTINGS COLLECTED/*BOUTURES COLLECTIONEES:*

clone no(s) -

Approximate no. of each clone/*No. approximatif de chaque clone* –

Date(s) of collection/*Date(s) de collecte* –

Where propagated/*Lière de multiplication* –

Rooting medium/*Milieu d'enracinement* – Length/*Longeur* – cm. +/- Auxin(*e*)

Number rooted/*Nombre enraciné* - on/*le* - (date) (%)

Number potted/*Nombre mis en pot* - on/*le* - (date) (%)

USE/*UTILISATION:*

No. planted for stockplants/*No. pour le parc à bois* – on/*le* – (date)

No. planted for clonal trials/*No. pour essais clonaux* – on/*le* – (date)

No. sent/*No. expédié* – to/*à* – on/*le* – (date)

(Example of a bilingual record sheet used in Cameroun)

This page has been left blank for field notes.

Check-lists, sources and records — *record sheet for propagation checks* A 68

Weekly checks:- **Week starting:** **Month:** **Year:**

Shading intact?
Polythene intact?

Water level? Rotting?

Rooting starting? Disease?

Ready to pot up? Pests?

Daily checks:-	Monday	Tuesday	Wednesday	Thursday	Friday	Saturday	Sunday
Conditions:							
Cuttings moist?							
Rooting medium moist?							
Air temperature?							
Soil temperature?							
Appearance of cuttings:							
Leaves falling?							
Wilting seen?							

Problems:-

What problem was:

Solution found?

How long problem lasted? What else needs attention?

This page has been left blank for field notes.